Fish

Fish

An Enthusiast's Guide

Peter B. Moyle

Illustrations by
Chris Mari van Dyck

UNIVERSITY OF CALIFORNIA PRESS
Berkeley · Los Angeles · London

University of California Press
Berkeley and Los Angeles, California

University of California Press
London, England

Copyright © 1993 by The Regents of the University of California

Line drawings and color illustrations Copyright © 1993 by Chris Mari van Dyck

First Paperback Printing 1995

Library of Congress Cataloging-in-Publication Data

Moyle, Peter B.
    Fish : an enthusiast's guide / by Peter B. Moyle ; illustrations by Chris Mari van
Dyck.

    p.   cm.
    Includes bibliographical references and index.
    ISBN 0-520-20165-5
    1. Fishes.   I. van Dyck, Chris Mari.   II. Title.
QL615.M63   1993
597—dc20   92-4806
                                                                                                   CIP

Printed in the United States of America
1   2   3   4   5   6   7   8   9

The paper used in this publication meets the minimum requirements of American
National Standard for Information Sciences—Permanence of Paper for Printed
Library Materials, ANSI Z39.48 – 1984. ⊗

Contents

Foreword

In this book, Peter Moyle successfully illustrates the joys of the study of living fishes, revealing why those of us who have spent a lifetime studying fish as a profession (ichthyology) consider ourselves to be so fortunate. We are constantly rewarded by discovering new and unexpected things that fish will do. Every visit to a stream or lake seems to add new insight into how fishes cope with their environment. These insights can be esoteric, such as discovering the sounds made by feeding pupfish in a crystalline Nevada spring. They can also be of considerable significance, such as discovering how different species recover from natural disasters like the "red tide" that killed most of the fish in 150 kilometers of the Pecos River in Texas, in 1986.

Unfortunately, those of us in ichthyology seem to be spending more and more of our time saving species from extinction and protecting aquatic habitats; these are important life-saving activities, and they are areas where amateur naturalists can play an especially important role. For example, I obtained considerable satisfaction from my role in saving a small fish species, the Big Bend gambusia, from extinction. At one time I had to drive five hundred miles with the entire population of this species in a container in my car: I drove very carefully. After the habitat in Big Bend National Park had once again been made suitable for them, the three surviving individuals were returned, and the species has since flourished.

The Big Bend gambusia is but one of many fish species whose survival is threatened by poor management of our increasingly scarce water supplies. All life needs water, and its scarcity is the ultimate ecological problem, as the steady expansion of the Sahara Desert demonstrates dramatically. Closer to home, the complete diversion of the Colorado River has resulted in the near elimination of its estuary. As a result, the totoaba, a fish species that once supported a thriving commercial fishery, is endangered because it requires the estuary for spawning. Declines of fishes like totoaba are good indications of excessive water withdrawal and symptoms of much larger problems. Fish diversity and abundance are also good indicators of water quality. Fish are much more likely to thrive in Clear Creek than in Salt Fork or Muddy Branch! Because Clear Creek contains the water we humans need as well, protecting the fish within it is usually in our own best interests. Perhaps the readers of this book will be stimulated to become guardians of their own Clear Creeks, if for no other reason than to have a place to watch fish.

Those of us who make a living studying fish also have another good reason for encouraging amateur naturalists: we benefit directly from their observations. An individual who has spent hundreds of hours fishing for largemouth bass in a particular lake often has extremely good insights into the biology of those bass, information that can be useful for management. Similarly, an aquarist who successfully breeds fish in captivity may possess unique knowledge that can give professional insights into the biology of poorly known species or even how to breed endangered species. In return, the professional may provide new perspectives for the amateur. For example, for many years I studied darters, which are small, colorful stream fishes. I was especially interested in the ability of different species to breed with one another and to produce hybrid young. An aesthetically pleasing result of these studies was the discovery that the offspring took on the color characteristics of both parent species. They were literally twice as colorful as a consequence. They were also sterile!

I like to think that ichthyologists enjoy special benefits available to few of our colleagues. We not only work with fascinating organisms, but we find that there are many nonprofessionals, amateur naturalists, if you will, who are as interested in our subjects as we are. Read this book and join us.

Clark Hubbs
University of Texas, Austin

Preface

To humans, fish are alien creatures. A fish lives suspended weightlessly in murky water, sensing the slightest movement in ways we can scarcely comprehend. Perhaps this is why fish swimming about in an aquarium can be so endlessly fascinating or why an angler can derive such pleasure just from being on the water, even if few fish are caught. It is certainly one of the main reasons the study of fishes has gained such favor in recent years with scientists and amateur naturalists. Nowhere among the vertebrate groups can you find such a diversity of body forms and colors, so many peculiar ways of making a living, or even so many bizarre (by our standards) ways of reproducing.

For example, the parrotfishes are large, brilliantly iridescent inhabitants of tropical reefs (see opening illustration, chapter 1). They make a living by using their powerful beaks to scrape algae and mucus from living coral. At night they may wrap themselves in blankets of slime after seeking refuge in a reef crevice. Mating can take place on a daily basis and often involves several different kinds of males, of different sizes and colors. If a parrotfish spawning congregation should lose, to a passing shark, the large male that normally dominates it, the largest female promptly changes sex and takes his place!

In the not-too-distant past, such fascinating tidbits of natural history were largely hidden beneath layers of water; knowledge of fishes was based mostly on what could be deduced from examining dead individuals, well removed from their natural surroundings. As a result, the different males and sexes of parrotfishes, with their various sizes and colors, were often described as separate species. We have more accurate information today because the development of scuba diving and snorkeling has allowed naturalists to observe fish in their own environment. The development of cheap but effective equipment for keeping fish alive in aquaria has also permitted a more intimate knowledge of their habits, and the large variety of fishes available in pet stores has increased awareness of their diversity. Accompanying these technological advancements has been a general increase in the interest in the natural history of fishes. This is reflected in the plethora of articles on fish found in the many specialized magazines now available to amateur aquarists and anglers, as well as in magazines with a broad natural history orientation. These articles in turn reflect the growing number of professional biologists, like myself, who choose fish as the focus of their studies, often aided by amateur naturalists. The purpose of this book is to make some of this fascinating information available to fish-watchers, aquarists, anglers, and other naturalists who share my enthusiasm for things ichthyological. It is my great hope that by providing this information I can also further the cause of fish conservation: many fish habitats and fish species will be lost forever if immediate action is not taken. How can our descendants fully comprehend the nature of the watery world if many of its most beautiful and mysterious inhabitants are gone?

AUTHOR'S ACKNOWLEDGMENTS

Much of the information in this book was extracted from *Fishes: An Introduction to Ichthyology* (2nd edition, 1988, Englewood Cliffs, N.J.: Prentice-Hall) by myself and Joseph J. Cech, Jr., so Joe Cech, the members of the University of California, Davis, Fish Ecology Research Collective, and the many other people acknowledged there made a major contribution to this book as well. I am grateful to Phil Pister for his comments on the conservation chapter and for being an inspiration to fish conservationists worldwide. It was a pleasure working with Chris Mari van Dyck because I could leave all the problems of developing illustrations for the book in her hands. Marilyn A. Moyle encouraged the project and patiently read over the manuscript with an English teacher's critical eye for spelling and grammar. Her comment that a previous draft of the book was boring and impersonal led to a major revision that resulted in a much more readable text. I dedicate this book to her and to our children, Petrea and Noah, who I hope will take some of the book's conservation messages to heart.

ARTIST'S ACKNOWLEDGMENTS

First, I thank Peter B. Moyle for the opportunity to be part of this project. His encouragement and patience along with that of Marilyn A. Moyle sustained me throughout.

I am very grateful for the professionalism, receptiveness, and support shown by the University of California Press staff. They made production a complete pleasure.

Much of my work would not have been possible without the tremendous contributions of illustrators, photographers, museum and aquarium personnel, and scientists before me. I have tried to specifically acknowledge these at the end of the book. I extend a sincere thank-you to all I have inadvertently overlooked.

I am indebted to my Mangan family, especially my husband Robert for his sense of humor and loyalty. My daughter Niall, whose beginnings coincided with the book's, and son Corey, whose beginnings entangled with its completion, broadened my perspective and deepened my purpose. I dedicate my part of this book in memory of my mother, T. Madalene MacDougall. Many of her dreams are reflected here.

CHAPTER I

Introduction

In a cool curving world he lies
And ripples with dark ecstasies.
   —*Rupert Brooke, "The Fish"*

The basic shape of a fish is simple; it can be drawn with a single sweep of a pencil. This elegant design is a reflection of how superbly adapted fish are to the "cool curving" world of water. To understand the ways of fish, the nature of their environment must be understood, as in many ways it is alien to us terrestrial creatures.

Because water is nearly eight hundred times more dense than air it is easy for fish to live suspended effortlessly in it, simply by balancing the heavy mass of bone and muscle with an internal float full of gas or lightweight oil. This means that their muscles can be devoted to producing the power it takes to push forward through the water, resulting in the streamlined shape that means "fish." A powerful, streamlined body is also necessary, however, because water resists movement much more than air—as any swimmer can attest. An extremely high proportion of each fish's body is devoted to swimming muscles. This is very convenient for those of us who eat fish, because these muscles make up the fillets that can be so easily removed with a few slices of a knife.

When pushing their way through the water, fish cause the water to swirl about them. The eddies created by the movement persist for some time after a fish creates them. Eddies are also created by the movement of the water itself, most noticeably as ocean currents, waves, and flowing streams. Thus the turbulence of water is a major environmental feature, much like wind over land. Fish have a special sensory system to detect this turbulence, which has no counterpart in mammals, birds, and reptiles. This **lateral line system** can provide many clues as to the

1

Figure 1-1.   Large males (*top*) of bluechin parrotfish usually result from a female (*bottom*) changing sex when a previously dominant male dies.

nearness of predators, prey, or school mates or to the presence of favorable or unfavorable environmental conditions. It is usually most visible as one or more narrow lines running down the side of a fish. In some fishes, the lateral line system has been partially converted into another sensory system that is equally alien to our experience: an **electrical sensory system**. This system takes advantage of the fact that water is a good conductor of electricity. In sharks, the electrical sense is used to detect the slight electrical fields generated by the muscles of their prey, whereas in some freshwater fishes it is used to monitor a special electrical field the fish set up about themselves. Objects and prey are identified by their differing abilities to conduct electricity and by how much they distort the electrical field around the fish producing it.

Other senses of fish are less alien to our experience, although water does put special constraints on them. Vision, for example, is quite important for most fishes, but its usefulness is often limited by lack of water clarity and by the way water acts as a selective filter of the spectral colors. Red light is excluded by the surface layers of water, whereas blue light penetrates the farthest. Thus the brilliant red fish frequently found in the ocean depths are nearly invisible in their natural habitat! Some fishes reduce their dependency on external light by producing their own light in **photophores**, whereas others, like catfish, rely on their sense of smell or taste to find their way about, following odor trails in the water or tasting and touching the bottom with sensitive whisker-like appendages (**barbels**).

Although water is a barrier to light, it is an excellent carrier of sound waves. In fact sound travels over three times faster in water than in air and carries much farther. Not surprisingly, most fish have an excellent sense of hearing, a fact we often do not appreciate because fish lack the external ears so characteristic of land vertebrates. They do not need external ears because the density of fish flesh is so close to the density of water that it carries sound waves with little distortion. The fish consequently need only an internal organ that is either more dense or less dense than water to intercept the sound waves and transmit the message to the inner ear. Usually this is done either by the air-filled swimbladder or by special earstones (**otoliths**). Many fish also produce sounds important for communicating with their fellows, especially when courting. The song of a courting toadfish may approach the hundred-decibel level! We rarely hear such sounds because sound waves do not move easily between water and air.

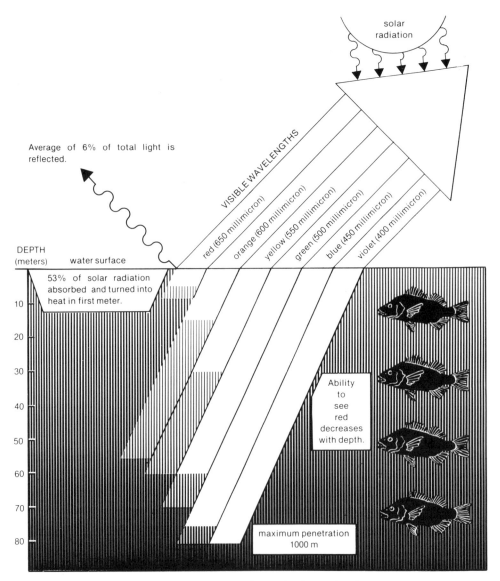

Figure 1-2.   Water acts as a selective filter of light. Red light waves penetrate the shortest distance, so a red fish that appears brilliantly colored when brought to the surface is in fact nearly invisible when below the level that red light penetrates.

Water carries sound waves so well because it is virtually incompressible. This same feature is used by many fishes to great advantage when they feed, because it causes water to rush into any new space, without any expansion. This is unlike air, whose gasses expand to fill empty space and compress easily as well. Most fishes are capable of rapidly increasing the size of the mouth cavity, forcing water to flow in rapidly through the small mouth opening, like sucking water through a straw. Fish that feed on insects and other small organisms can thus literally suck their prey into their mouths. The water taken in is expelled through the gills by closing the mouth and compressing the mouth cavity. The food is retained by the **gill rakers**. These projections from the supporting arches of the gills function much like the bars of a cage. The smaller the prey, the closer together the gill rakers. This expandable mouth cavity is also very handy for breathing (respiration) because it allows large volumes of water to move continuously across the gills, even when the fish is not feeding.

Having gills that are efficient at extracting oxygen from the water is important because water typically contains less than 8 milliliters of oxygen per liter of water, compared to 210 milliliters in a liter of air. Furthermore, the capacity of water to hold oxygen decreases as temperature increases, while at the same time the fish's demand for oxygen is increasing. The decay of organic matter, either natural or man-made, also removes oxygen from the water. Thus the activities of fish often may be limited by the shortage of oxygen in the water, even though the gills may be extracting most of the oxygen available. The efficiency of the gills depends on having a vast surface area in the gill filaments and a multitude of blood vessels into which the oxygen can be taken from the passing water. The fish take advantage of proximity of the blood to water by using the gills to eliminate waste carbon dioxide, ammonia, and heat at the same time that they take up oxygen. This exchange system also creates some problems, however, because it makes fish very vulnerable to other substances dissolved in the water, especially pollutants such as mercury or pesticides which can kill the fish after being taken up through the gills. The thin membranes that separate the blood from water can act as selective filters to some compounds, however, especially salts. Few fish can live in both fresh and salt water. Marine fishes need to filter out excess salt. Freshwater fishes have the opposite problem of needing to retain salt, because there is so little salt in fresh water.

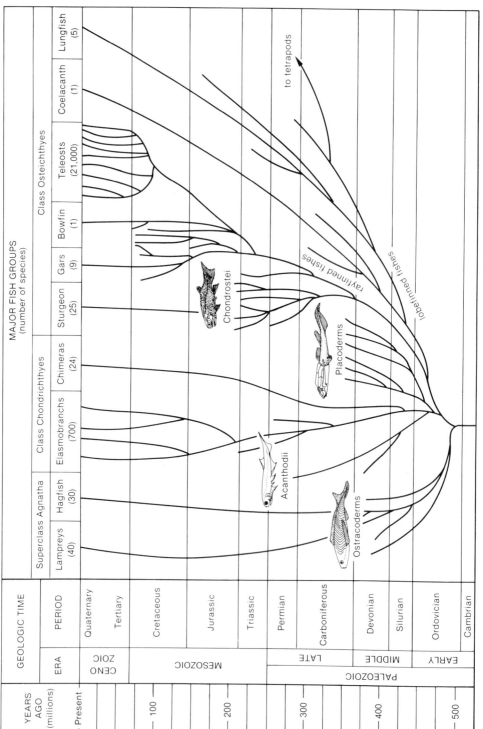

Figure 1-3. An evolutionary tree for fish and agnathans.

EVOLUTION AND CLASSIFICATION OF FISHES

The history of fishes through time is fascinating not only because it helps to explain the development of their many exquisite adaptations but because the early history of fishes is that of all vertebrates, including ourselves. Fish made their first appearance in the fossil record nearly 500 million years ago. These early vertebrates, called ostracoderms, were small, heavily armored forms that seem to have lived by sucking algae, small invertebrates, and ooze from the bottoms of seas and lakes. They lacked both the jaws and paired fins that are characteristic of "higher" fishes. Their closest relatives alive today are lampreys (order Petromyzontiformes) and hagfishes (order Myxiniformes), strange eel-like creatures that also lack jaws and paired fins. Hagfishes are so peculiar looking that early taxonomists classified them as worms! Despite their ancient affinities, both hagfishes and lampreys are quite abundant today, in part because they can prey on more "advanced" fishes.

The first fishes with jaws and paired fins to appear as fossils, some 440 million years ago, were the acanthodians, small minnow-like creatures. These fishes had a whole row of fins on each side of the body, unlike more modern fishes that have only two sets of paired fins (the equivalent of legs on land vertebrates). The relationship of the acanthodians to modern fishes is a subject for debate, as is the relationship of their contemporaries, the placoderms (class Placodermi). This bizarre collection of armored fishes greatly overshadows the acanthodians in the fossil record. The placoderms were mostly bottom-dwellers; some were fearsome predators 10 or more meters in length. They dominated the oceans for about 100 million years but were replaced completely by the two groups that dominate the waters of the world today, the class Chondrichthyes (cartilaginous fishes, such as sharks and skates) and the class Osteichthyes (bony fishes). The sharks and skates have had almost their entire evolutionary history in salt water, in contrast to the bony fishes which seem to have developed initially in fresh water and invaded salt water fairly late in their evolution. As a consequence of their independent evolutionary histories, the two groups have developed rather different solutions to the problems of living successfully in water. The sharks, for example, store lightweight oils in their livers to make themselves buoyant, whereas bony fishes use gas bladders. Other differences are discussed in chapter 5.

Early in their evolutionary history, the cartilaginous fishes split into two independent lines: the sharks, skates, and rays (subclass Elasmo-branchi)* and the ratfishes (subclass Holocephali). The sharks, skates, and rays only number about seven hundred species but they are quite successful and widely distributed as top predators. Some species even feed largely on mammals and birds and a few others have invaded fresh water. The ratfishes are a rather peculiar group of about thirty bottom-dwelling species, named for their long, rat-like tails. They have changed little since they first appeared in the fossil record, being specialized for crunching clams and other invertebrates in the ocean's depths.

Like the cartilaginous fishes, the bony fishes split into different evo-lutionary lines early in their history. One line, the ray-finned fishes (sub-class Actinopterygii), gave rise to the bony fishes that dominate the waters today. The three other lines resulted in small, rather obscure groups that all possess lobed (limb-like) fins. These obscure fishes are of considerable interest, however, because one of the lines gave rise to the terrestrial vertebrates (tetrapods). One of the more prominent living groups, the lungfishes (subclass Dipneusti), even breathe air. Another group, the coelacanths (subclass Crossopterygii), are most famous be-cause they were thought to be extinct for nearly 10 million years until a fisherman pulled one from the depths of the ocean off the coast of Africa. It is perhaps the only fish that has made newspaper headlines all over the world.

Although the cartilaginous and lobe-finned fishes are of great interest, their role in this book is minor compared with that of the ray-finned fishes. The rayfins, with over 21,000 species, dominate both fresh and salt water in both number of species and number of individuals. The first fossils appear in freshwater deposits over 400 million years old. By about 340 million years ago they had assumed their dominant position in fresh water and were moving into the seas, eventually to displace the placoderms. The earliest ray-finned fishes were the chondrosteans (in-fraclass Chondrostei), represented in the modern fauna only by about twenty-five species of sturgeon and two species of paddlefish (order Acipenseriformes). They were gradually replaced by various groups of more "modern" fishes (infraclass Neopterygii), represented today by the gars (order Lepisosteiformes, nine species) and the bowfin (order Ami-iformes, one species). The gars, bowfins, and their relatives dominated the seas and fresh waters in the age of dinosaurs but were replaced in their turn by the dominant modern fishes, the teleosts (subdivision Te-leostei). The teleosts have diversified in many directions, producing forms adapted for living in most of the aquatic habitats of the world.

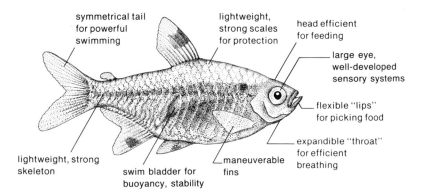

Figure 1-4. Some traits of a successful teleost, as illustrated by a small, nearly transparent tetra that is commonly kept in aquaria.

The many evolutionary directions are represented by the twenty-five to thirty teleost orders (the number depends on who is doing the classifying), the hundreds of families, and thousands of species that have been described.

Why have the teleosts been so successful? Their success is due to a set of adaptations that together have made this group superbly adapted for living in water. The tail is symmetrical and provides even, powerful thrust for swimming. The scales are thin but strong, providing protection without weighing down the fish with armor. The skeleton is mostly bone, with strong but light construction. The skull is complex in structure and has been modified for many specialized types of feeding, especially suction feeding. The gas bladder is used to create neutral buoyancy and its volume can be finely controlled, making it possible to move through the depths. The fins are highly maneuverable and give each fish fine control over its movements. The body shape shows tremendous variability (see chapter 2) so the fishes can occupy many unusual habitats (especially the many small species).

HABITATS OF FISHES

The amazing array of fish body shapes and sizes reflects the equally amazing array of habitats in which they can be found. There are fish in mountain lakes above 4,000 meters and fish in the deep ocean, at depths of at least 8,300 meters. In the outflows of hot springs fish can be found living at temperatures of around 42° C, whereas in the Antarctic, fish can be found resting on chunks of ice surrounded by water that is within a fraction of a degree of freezing solid. The salinities of the water in which fish are known to live range from the purest water of granitic mountain basins to water over four times saltier than sea water. The oxygen content of the water can be at saturation, as in cold mountain streams, or at essentially zero, as in tropical swamps (where most fishes breathe air). Even light does not seem to be necessary for fish life. Some of the most peculiar-looking fishes are those inhabiting the lightless depths of the ocean or deep underground aquifers. Socially, fish can be found in densely populated coral reefs where hundreds of individuals representing dozens of species can be found in areas only a few meters square, or they can be found singly, sparsely scattered in mountain lakes or the waters of the open ocean.

Despite their enormous ability to adapt to a wide range of aquatic environments, fish are far from being uniformly distributed in the waters of the globe. Thus 41 percent of all fish species are found exclusively in fresh water, even though fresh water covers only about 1 percent of the earth's surface and comprises less than .01 percent of the earth's water by volume. The reason for this is that fresh water is a highly diverse and fragmented habitat, so that fish populations frequently become isolated from one another, an important condition for the evolution of species to take place. Isolation is also promoted by the instability of the earth's crust and climates, as mountain ranges rise and fall and glaciers advance and retreat. The oceans, of course, are major barriers to the movements of freshwater fishes; only 1 percent of fish species move freely between the two environments. Perhaps because so few species are capable of this feat, the species that do move between the environments are often very abundant: salmon, shad, eels, striped bass.

Although 97 percent of the earth's water is in the oceans (the 3 percent remaining consists of 1 percent in fresh water and 2 percent in glaciers and the atmosphere), most of this water is open at best to species highly specialized to live under great pressure and food scarcity. The mean depth of the ocean is around 4,000 meters and 98 percent of its water is below 100 meters, the maximum depth at which enough light can penetrate to allow plants to grow. Seventy-five percent is below 1,000 meters, the usual total limit to light penetration. What all these numbers mean is that most marine fishes live in about 2 percent of the available water, although the deep ocean is so voluminous that about 12 percent of all known species are found there. Even the 2 percent figure for prime fish habitat, however, is deceptive, because much of that

Figure 1-5.   The importance of fish to humans is well
illustrated by this design from an ancient Mimbres bowl
from arid central New Mexico. Mimbres people traveled
long distances and recorded not only local fish but fishes
from the reefs in the Gulf of California.

water is in the vast surface reaches of the open ocean, which are turbulent, featureless, and generally unproductive. They support as a consequence only about 1 percent of the fish species. This leaves only the
narrow strip of shallow water along the continents, above the continental shelves, and the reefs around oceanic islands to support 44 percent
of all fish species and the majority of individual fish as well. The size of
our marine fisheries is an indication of the richness of these areas: 70
million metric tons of fish are taken from the sea each year, mostly close
to the continents in temperate areas. Even more remarkably, over 30
percent of all known fish species are found in direct association with
tropical reefs, which occupy just a tiny portion of the marine environment. This book can only sample the diversity of fishes and aquatic
environments, but it should be viewed as an invitation to the reader to
explore them further, preferably by getting wet.

VALUE OF FISH

Humans are increasingly dependent upon fish for protein. If we assume
a world population of 5 billion people (headed for 10) and, conserva-

tively, an annual catch of fish from all sources (including fish farming) of 100 million metric tons, roughly 20 kilograms (44 pounds) of fish are caught and consumed per person each year. The actual amount is probably somewhat higher. The monetary value of world fisheries is hard to estimate because there are so many poorly recorded fisheries. However, an indication of the value is given by the fact that the United States alone imports about $3 billion worth of fish products each year, and American anglers spend over $8 billion each year on their hobby. Even the trade in aquarium fishes has become a major worldwide industry, valued at hundreds of millions of dollars per year. The value of the aquarium trade is indicative of the increasing awareness of the aesthetic value of fish. This is seen also in the increasing numbers of people taking up scuba diving and snorkeling as hobbies and, of course, in the increasing number of intelligent people reading books like this one.

FISH CONSERVATION

The ever-increasing popularity of fish has its down side: commercial fish stocks are being depleted everywhere, from the once phenomenally abundant herring and cod to the rare and delicate species sought by the aquarium trade. Equally alarming is the decline of fish diversity as we divert and pollute the fresh waters of the world or dump our wastes into productive coastal waters. In regions of the world with Mediterranean climates, where humans love to live but fresh water is scarce (such as California, South Africa, Chile, or Spain), 60 to 70 percent of the native freshwater fishes are well on their way to extinction. Some species are already gone. In tropical areas, fishes are being lost faster than they can be described by ichthyologists—probably two hundred species of cichlids have been lost from Africa's Lake Victoria alone (see chapter 15). Thus one of the goals of this book is to create a greater appreciation for what we are losing in hope that greater awareness can at least slow down the rate of loss of our beautiful, irreplaceable fishes.

CHAPTER II

Fish from the Outside

It is said that when the great nineteenth-century biologist Louis Agassiz took on a new student, his first act was to lock the student in a room for a day, with only a dead fish for company. At the end of the day the student reported to the professor all he had learned about the fish from his examination of its features. Although this procedure is no longer standard practice at our universities, it is still true that you can learn a great deal about the biology of a fish simply by looking closely at its external anatomy. The external features of a fish can provide clues as to where a fish lives and how it makes a living. The purpose of this chapter is to present enough background material so the reader will be able to make intelligent guesses about a fish's place in the environment from a quick examination of its external features.

VOCABULARY

When Dr. Agassiz's student was let out of his room, fish in hand, he would have found it much easier to discuss the fish with the learned professor if he knew the names of its various parts. Likewise, you will find this book much more enjoyable if you learn some basic vocabulary.

Examine carefully the labeled drawings of the shark, the trout, and the snapper. Note first that there are two sets of paired fins, equivalent to limbs of terrestrial vertebrates: the **pectoral fins** and the **pelvic fins**. The pectorals are usually in front of the pelvics. On top of the fish is the **dorsal fin**, which is divided in some species into two or three sep-

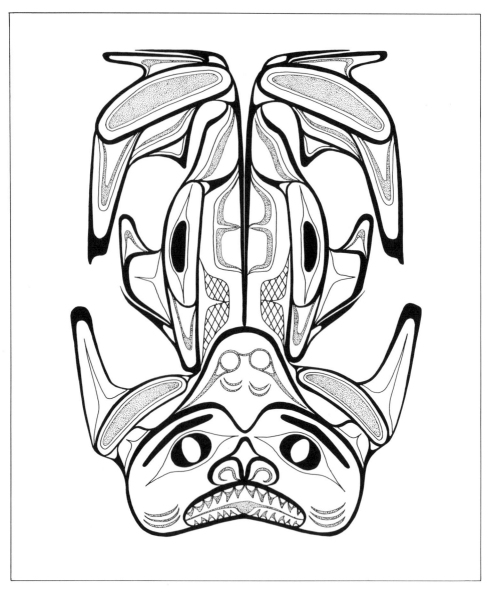

Figure 2-1. A dogfish shark as depicted by a Haida artist of the Pacific North-west. All the external features found in figure 2-2 are also present on this fish, but they are artistically rather than accurately (by our standards) arranged.

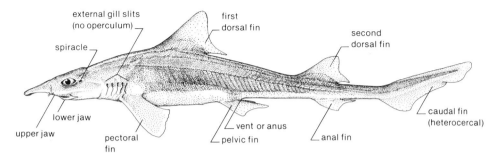

Figure 2-2.   External features of a shark, the brown smoothhound. The absence of claspers on the anal fin indicates that this fish is female.

arate parts or, in the case of sharks, may be two separate fins entirely. The **adipose fin**, shown on the trout, is a small fin behind the dorsal that is characteristic of a number of important fish groups, such as salmon and trout, catfishes, and characins (tetras, etc.) but is absent from most fish. The **anal fin** begins just behind the vent or anus. The **caudal fin** is the tail and is supported by the region of the body called the **caudal peduncle**. The principal feature to note on the body itself, in addition to scales, is the **lateral line**, which is part of a special sensory system and appears as a pencil-thin line of pores on each side of the fish. The **operculum** is the cover for the gills; it is connected to the **branchiostegal rays** on the bottom of the head. These rays are a fan-like arrangement of bones that help pump water across the gills. In shark-like fishes, the operculum is absent and in its place are a series of **gill slits** and a small round opening, the **spiracle**. In bony fishes, the chief bone of the lower jaw is the **dentary** (lower mandible), whereas the upper jaw usually consists of two bones, the **premaxilla** and the **maxilla**. The premaxilla is usually the bone that forms most of the extendable "upper lip" of many bony fishes.

The trout diagram shows three different ways of measuring the length of a fish. **Total length** is the distance from the snout to the end of the tail; it is preferred by anglers because it provides the maximum length of a fish, useful for bragging. The problem with total length is that it is quite variable because tail shapes vary and tail tips are often frayed or worn. **Fork length,** the distance from the inside fork of the tail to the snout, is more reliable and is widely used by fisheries biologists because it can be measured rapidly. **Standard length** is the measurement pre-

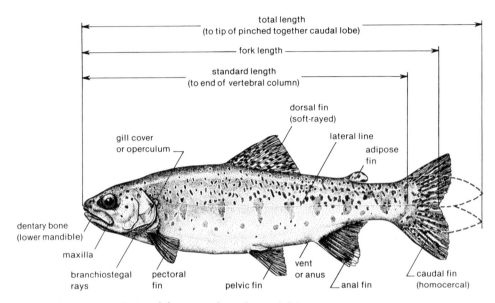

Figure 2-3.   External features of a soft-rayed fish, the rainbow trout, showing different ways of measuring the length of a fish. Total length is preferred by anglers; the other two are used by biologists because they are less subject to argument.

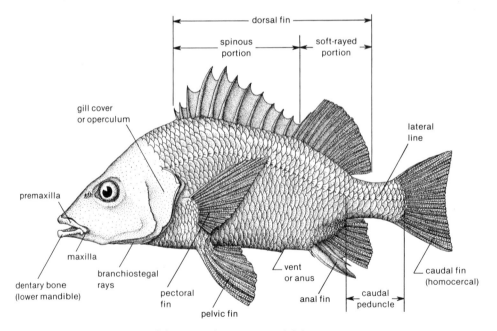

Figure 2-4.   External features of a spiny-rayed fish, a snapper.

ferred by fish biologists, because it can be compared among species and can be made on specimens that have been preserved for a long time and may have lost or damaged their tails. This is the distance from the tip of the snout to the end of the vertebral column, which can generally be seen as a ridge across the end of the caudal peduncle when the tail is bent.

BODY SHAPE

Fishes can be arbitrarily divided into six types, according to body shape, with each type reflecting a distinct way of life: rover-predator, lie-in-wait predator, surface-oriented fish, bottom fish, deep-bodied fish, and eel-like fish.

Rover-predators have the body shape that is most generally associated with the general concept of "fish." Their bodies are smooth and streamlined, with a pointed head, narrow caudal peduncle, and forked tail. The dorsal and anal fins are well developed, and the paired fins are all about the same size. The eyes are fairly large, as is the mouth, which is located at the end of the snout. These fishes are well adapted for cruising about and capturing any suitable prey they see with a sudden burst of speed. Some of the better known examples include tuna, mackerel, bass, swordfish, requiem sharks, and many species of minnows. Trout and other stream fishes possess this body shape because it is an efficient one for swimming through fast-moving water.

Lie-in-wait predators also capture their prey with a sudden burst of speed, but they typically do this from ambush. Their body is torpedo-shaped, with a large caudal fin. The dorsal and anal fins are placed far back on the body. The pectoral fins are usually enlarged and placed close to the head, which is flattened. The mouth is large and contains numerous sharp, pointed teeth. The placement of the dorsal and anal fins allows them to assist the caudal fin in producing the thrust needed for a fast start. The long narrow body and flat head may cause the prey (usually other fish) to misjudge the distance separating them from the predator, as well as the size of the predator, until it is too late. Although the classic examples of lie-in-wait predators are the true pikes, this body shape has evolved independently in many other groups: barracudas, trumpet fishes, squawfishes, even pike-killifishes.

Surface-oriented fishes are small in size, with upward-pointing mouths, heads flattened on top, large eyes, and dorsal fins placed towards the rear of the body. Fishes with this body shape are particularly

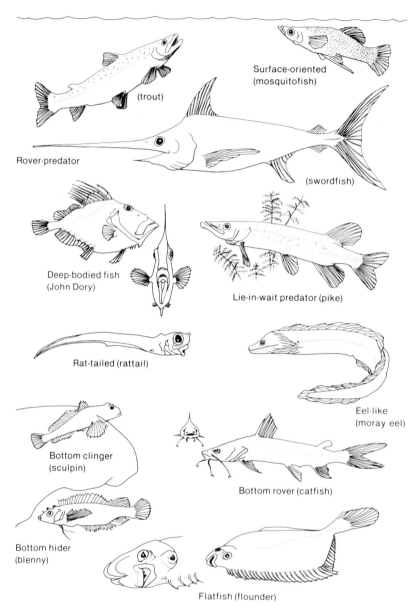

(trout)

Surface-oriented
(mosquitofish)

Rover-predator

(swordfish)

Deep-bodied fish
(John Dory)

Lie-in-wait predator (pike)

Rat-tailed (rattail)

Eel-like
(moray eel)

Bottom clinger
(sculpin)

Bottom rover (catfish)

Bottom hider
(blenny)

Flatfish (flounder)

Figure 2-5.   Types of body shapes in fishes.

well adapted for capturing insects that live on or close to the surface of the water (e.g., emerging mosquitoes) and for surviving in stagnant water. The latter is possible because they can align their flattened heads with the surface of the water and suck in, through their upward-pointing mouths, a mixture of air and water. This oxygen-rich mixture is pumped across their gills. Most surface-oriented fishes, such as mosquitofish, guppies, and many killifishes, occur in fresh water, but some marine forms, such as halfbeaks and flying fishes, also possess a variant of this body shape.

Bottom fishes actually come in a variety of shapes, but all are adapted for living in close contact with the bottom. Most are flattened in one direction or another. They can be divided into five body types that broadly overlap with one another: bottom rovers, bottom clingers, bottom hiders, flatfishes, and rattails. **Bottom rovers** resemble rover-predators in body shape but their heads tend to be flattened, their backs humped, their eyes small, and their pectoral fins enlarged. Many (but not all) have their mouths located beneath their snouts (subterminal). The many species and families of catfishes mostly fall in this category, as do carp, suckers, and sturgeon. The latter fishes all have fleshy, extendable lips that can be used for sucking food off the bottom. Extendable lips are also characteristic of the many small shark species that fall into this category, with their subterminal mouths, flattened heads, and large pectoral fins. Some of these sharks feed by biting off the protruding siphons of otherwise buried clams.

The next type of bottom fish, **bottom clingers**, are small fishes with large flat heads and pectoral fins. Their pelvic fins are modified in ways that allow them to hold on to the bottom, an advantage in areas where currents are strong. The most obvious examples are the clingfishes, gobies, and hillstream fishes, all of which have their pelvic fins fused together to form a suction cup. A much simpler device is found in sculpins, small fishes common in streams and tide pools; their pelvic fins are small, straight, and closely spaced, forming an anti-skid plate that provides plenty of traction for holding on to the bottom. **Bottom hiders** are also small fishes with large pectoral fins, but they lack clinging devices and are more elongate, with smaller heads. They live under rocks and in crevices, often in still water. Examples include the darters that are common in streams of eastern North America and many of the tropical marine blennies.

The most extraordinary of the bottom fishes are the **flatfishes**, which include flounders and soles among the bony fishes, and skates and rays among the cartilaginous fishes. Flounders and soles are actually deep-bodied fishes (see next category) that have evolved to live with one side on the bottom; during their development, the eye that should be on the bottom side migrates to the top side and the mouth acquires a peculiar twist to permit picking up prey from the bottom. Skates and rays, although just as flat, do not have to go through such developmental contortions to achieve their body shape; they are essentially flattened sharks, with greatly enlarged pectoral fins and greatly reduced tails. They move about by flapping or undulating their pectoral fins. Their mouth is completely ventral, whereas the intake for their gills (the spiracle) is located on top of the head. The latter arrangement seems much more efficient than that of the flounders, which have to contend with having one of their gill openings on the bottom.

The **rattail shape** has also evolved independently in both bony and cartilaginous fishes, mainly among marine fishes that live in fairly deep water, such as grenadiers, brotulas, and chimaeras. These fishes have large, pointy-snouted heads and large pectoral fins but the body quickly tapers down to a long pointed tail. Exactly why this body type is popular among deep sea fishes is a bit of a mystery, but it must be advantageous in their lives as scavengers and predators on bottom-dwelling invertebrates.

Deep-bodied fishes are flattened from side to side and have a body depth that is at least one-third the body length. Most have spines in their fins and have long dorsal and anal fins; the pectoral fins are located high on the body with the pelvics in a supporting position below. The mouth is small and flexible, the eyes large, and the snout short. Such fishes usually specialize in picking small invertebrates from the bottom or small zooplankton from the water column, because the body shape affords a great deal of precision in movement. The spines are necessary for protection because these fishes have sacrificed speed for maneuverability; this also means they are rarely far from cover, so are usually found on reefs, in beds of aquatic plants, or near other structures. Examples include sunfishes, butterfly fishes, and cichlids, such as tilapia. Deep-bodied fishes that lack both spines and association with cover are usually plankton-feeding fishes such as herring or menhaden. Such fishes are really specialized rover-predators that have added a sharp keel to their belly. This keel eliminates the faint ventral shadow more rounded fishes possess, making the plankton-feeder less visible to predators approaching from below.

Eel-like fishes have extremely long bodies, blunt or wedge-shaped heads, and tapering or rounded tails. They usually possess long, low dorsal and anal fins, which may unite with the tail, but paired fins are optional. Scales, if present, are small and imbedded, so the skin appears smooth and slimy. Eel-like fishes live in such tight places as crevices and holes in reefs and rocky areas, beds of aquatic plants, and burrows in soft bottoms. The best examples are the over six hundred species of true eels, but the body shape is also found in loaches, gunnels, and a number of other fishes and is even approached in a few odd sharks.

SCALES

Scales are the principal body covering of fishes and so can provide many clues as to how a fish lives.

Types of scales.   The most primitive scales on modern bony fishes are the heavy **ganoid scales** of North American gars. These scales are hard and diamond shaped, forming a nearly impenetrable coat of armor. This armor is necessary for fish that spend most of their time lying in wait for passing fish to ambush or avoiding alligators, one of the few predators that can crush their armor. In contrast, most bony fishes possess thin, flat scales that provide some degree of protection yet allow the fish more speed or maneuverability than is possessed by gars. These scales are of two basic types: cycloid and ctenoid. **Cycloid scales** are the smooth scales that characterize the fishes that lack true spines such as herrings, trout, eels, and minnows. **Ctenoid scales** have tiny projections covering the outer edge of each scale; these give the bearers the rough feel associated with spiny-rayed fish such as perch or sunfish.

A surprising number of bony fishes have dispensed with scales altogether or have them so deeply imbedded in the skin they are difficult to detect; these fishes are either those that spend much of their lives hidden in tight places (such as eels, sculpins, and catfishes), or those that spend their lives swimming at high speed (such as tunas, swordfish, and mackerels). Curiously, many tropical catfishes have developed bony plates on their backs that function as armor, much as ganoid scales do in gar. Equally curious are the **placoid scales** of sharks, which are tiny and give sharkskin its sandpaper-like feel; the scales function much like the ctenoid scales of advanced bony fishes, but evolved millions of years earlier. Most skates and rays lack the covering of placoid scales, although a few are retained and enlarged for such functions as the "sting" on the tail of a stingray.

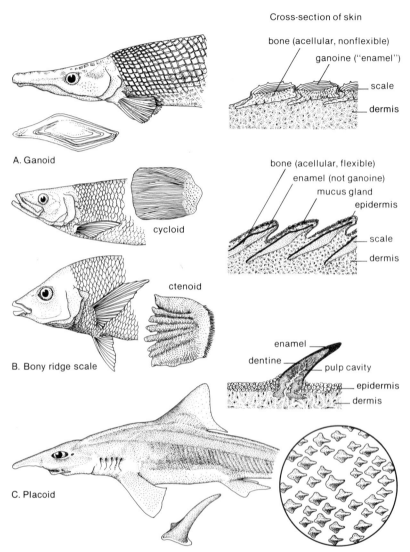

Cross-section of skin

bone (acellular, nonflexible)
ganoine ("enamel")
scale
dermis

A. Ganoid

cycloid

bone (acellular, flexible)
enamel (not ganoine)
mucus gland
epidermis
scale
dermis

ctenoid

B. Bony ridge scale

enamel
dentine
pulp cavity
epidermis
dermis

C. Placoid

Figure 2-6.   Types of fish scales. Ganoid scales are typical of only a few "ancient" fishes, such as gars; cycloid scales are typical of soft-rayed fishes, and ctenoid scales are typical of spiny-rayed fishes. Placoid scales are found only on sharks, skates, and rays and are remarkably similar in structure to teeth.

Number and size.   Fishes that possess large numbers of small scales usually spend most of their time cruising about (sharks, mackerel, many minnows) or else hold in fast currents (trout). Fishes with larger scales tend to be more sluggish and/or inhabit quiet water, presumably because the scales function in part as lightweight armor. One apparent exception to this "rule" is herring, which are constantly in motion yet possess fairly large scales. Their larger scales are advantageous because they are silvery, functioning as individual mirrors to reflect light and confuse predators. These scales are also shed easily, making it difficult for a predator to get a firm grip on its prey, so they may wind up with a mouth full of scales rather than a whole fish.

FINS

The shape and position of the fins are closely tied to body shape. Bottom fishes, for example, typically have large pectoral fins and rounded caudal fins, whereas eel-like fishes often lack one or both sets of paired fins completely and have pointed caudal fins. Still, the fins by themselves can tell much about a fish.

Paired fins, that is, the combination of pelvic and pectoral fins, are the main means most fish have for fine control of their movements. In fact, the evolution of paired fins in ancient fishes is considered to be one of the major steps in the evolution of the vertebrates for that reason. In general, most of the steering is accomplished by the pectoral fins, assisted by the pelvics, which act mainly as stabilizers to keep the fish from rolling from side to side. In rover-predator fishes, the two sets tend to be widely separated and located low on the body, providing stability while the fish is cruising or holding in current. In deep-bodied fishes, however, the pectoral fins are located just behind the operculum and fairly high on the body, close to the center of gravity, whereas the pelvics are placed in a supporting role below them. In such fishes, the pectoral fins are inserted vertically, rather than horizontally, often with a "wrist" that gives the fins great flexibility of movement. As a result, the fishes are capable of holding position while delicately sucking small invertebrates from plant stems or rocks.

Pectoral fins that are long and pointed permit very fine control over movement, enabling their possessors to feed on very tiny and elusive prey. Broad, rounded pectorals are found on bottom fishes and sluggish predators, although the largest pectoral fins are found on flying fishes and on fishes like the "flying" gunards that use their fins for displays.

Dorsal and anal fins tend to be long on deep-bodied fishes, for stability while swimming, and on eel-like fishes, where they assist the sinuous mode of swimming. In the peculiar African and South American electric fishes, and in seahorses, the dorsal (or anal) fin is the main means of propulsion, as the fishes must maintain rigid bodies. Short dorsal and anal fins are most characteristic of bottom and surface-oriented fishes, and of fishes, such as trout, considered to be less advanced than the spiny-rayed fishes (which typically have long dorsal fins).

Caudal fins have shapes that are strongly related to the cruising speed of fishes. The fastest fishes are tunas, swordfish, and mackerel sharks which have stiff, quarter-moon–shaped tails attached to a narrow caudal peduncle. This type of tail can be moved back and forth extremely rapidly. Rover-predators, and other fishes that require rapid or continuous movement, have forked tails; generally, the deeper the fork, the faster the fish can swim (within its taxonomic group). Thus among the catfishes, the channel catfish, with a deeply forked tail, can swim faster than the white catfish, with a moderately forked tail, which can swim faster than any of the bullhead catfishes, with rounded tails. Deep-bodied, surface-oriented, and most bottom fishes have square, rounded, or slightly forked tails. The tails of most bony fishes have the upper and lower lobes the same size, called **homocercal tails**, whereas most sharks have tails with the upper lobe much longer than the lower lobe, called **heterocercal tails**. Heterocercal tails are often considered to be more primitive because they evolved first, but the heterocercal tail of modern sharks is as advanced functionally as the homocercal tail of bony fishes.

Spines are important parts of fins and have evolved independently a number of times in fishes. "True" spines are found in the dorsal, anal, and paired fins of the perch-like fishes and are considered to be the hallmark of this group, which dominates shallow marine environments and many freshwater environments as well, particularly lakes. Spines that are in reality just single, stout, stiffened fin-rays are found on such fishes as catfish, carp, and goldfish. Many sharks and rays also have single spines preceding their dorsal and anal fins but they are derived from placoid scales. Spines are important because they are an effective, lightweight defense against predators, especially when they have mechanisms that allow them to lock into position and, in some fishes, have poison glands associated with them. They work not only because they are sharp but because once a fish has erected and locked in place the spines in its dorsal, anal, and pectoral fins, it has effectively increased its size as far as predators are concerned. A spiny fish is a much bigger mouthful than a fish without spines of the same body size and it is a fundamental rule in nature that the bigger you are, the fewer animals there are around to eat you.

Not surprisingly, spines tend to be best developed on small- to medium-sized fishes that feed in the open, such as sunfishes, catfishes, and butterfly fishes. Curiously enough, there are many fishes (e.g., sculpins, blennies, gobies) that have true spines in their fins, yet these spines are as soft and pliable as fin rays; these fishes are all small, highly camouflaged, and live under rocks and logs, where rigid spines would interfere with their ability to squeeze into tight places.

SENSE ORGANS

The most noticeable sense organs on fishes are usually the **eyes**. Their size and position can tell much about when and how a fish feeds. Fish that are active during the day have large eyes, relative to their body size, although the largest eyes usually belong to species that feed at dawn or dusk, or in deep water. For example, a freshwater fish with unusually large eyes is the walleye (or walleyed pike), which feeds at low light levels. It has a special light-gathering layer of tissue in the eye as well, called the **tapetum lucidum**, which increases its ability to see under poor light conditions. When a light is shone at a walleye, this layer makes the eyes seem to glow in response. Fish with small eyes are often nocturnal or live in dark or turbid places, where vision has limited usefulness. Many such fishes have "whiskers" around their mouths (e.g., catfish), called **barbels**, which are used to find prey on the bottom, through a combination of taste and feel.

The sense of smell may also be well developed in such fishes, although the nostrils on most fishes are not very conspicuous. The exceptions are the sharks, skates, and rays whose large nostrils have given them an undeserved reputation of being largely dependent upon smell to find their prey. However, one of the functions of the peculiar head of the hammerhead shark appears to be to keep the nostrils widely separated, providing a "stereoscopic" sense of smell. Odor trails can be followed by turning to the side in which the smell is strongest. If the importance of the sense of smell to sharks is overemphasized, the importance of hearing to all fishes is underemphasized. As pointed out in the last chapter, this is because fish lack external ears like land animals. Their hearing is nevertheless keen, especially for low-frequency sounds.

The sensory system of fishes for which we have the least empathy is the **lateral line system**, known technically as the acoustico-lateralis system. This system is generally visible as a single line on each side of a fish although not all fishes have a conspicuous lateral line and some have several. Branches of this system are also present on the head but are hard to see. The lateral line system detects changes in turbulence along a fish, as well as changes in pressure. It allows fish to swim about in the dark and to avoid running into things, such as the sides of an aquarium. In some fishes, such as sharks, parts of this system have evolved into sense organs capable of detecting electrical fields, useful for navigating long distances (by detecting changes in the earth's electrical fields) and for finding prey (by detecting the electrical field generated by the normal activity of muscles). Most sharks shut their eyes as they close in on their prey, guided instead by sensing the faint electrical field.

MOUTH

The mouth of a fish often provides obvious clues for determining how it makes a living. Fish with large mouths filled with sharp teeth, such as bluefish, barracuda, gar, and white sharks, obviously are predators on fish and other large soft-bodied animals. A large, hard-rimmed mouth is the best sign of this, because many piscivorous (fish-eating) fish have large sharp teeth only in the throat (**pharyngeal teeth**) or have patches of small teeth on the roof of the mouth. Squawfish, fallfish, and creek chub are examples of fishes that lack teeth in the mouth yet are formidable predators on small fish because they possess sharp pharyngeal teeth. Also fairly obvious in function are the downward pointing mouths of suckers, carp, and sturgeon with soft "lips" that appear remarkably humanoid. Such mouths function in sucking invertebrates, algae, and organic debris off the bottom.

Most bony fishes, however, have small- to medium-sized mouths, more or less on the end of the snout, with bony "lips." Such mouths are remarkably flexible and have been adapted for capturing a wide variety of prey. The bony lips (especially the premaxillary and maxillary bones of the upper jaw) can be protruded rapidly, to form a small round opening. When this happens, the mouth cavity is increased in size simultaneously, by the expansion of the branchiostegal rays. As a result, water rushes in with great speed, carrying with it whatever small organisms are in the stream. This manner of feeding, which mimics the action of a pipette or eyedropper, is one of the major reasons bony fishes are so abundant; it allows them to specialize on small organisms that would be difficult to capture by grabbing and biting with rigid jaws.

Generally, the smaller the mouth, the smaller the prey. Beyond this, there are many unusual specializations of fish mouths. For example, some species of African cichlids feed on the scales of other fishes. To remove each scale, they slide their flat lower jaw under it, clamp down with the flexible, toothed upper jaw, and jerk the scale out. Equally remarkable are the butterfly fishes that have tiny mouths lined with patches of tiny teeth, on the end of long snouts; they live by scraping small sponges off the inside of crevices in coral reefs.

GILL OPENINGS AND OTHER STRUCTURES

The gills of bony fishes are covered by the thin but flexible **opercula**. These bony gill covers not only protect the gills but are part of the respiratory pumping system that keeps a steady flow of well-aerated water moving across the gills. Thus, the shape and relative size of the gill openings on bony fishes tends to be fairly constant. They are reduced, however, in eels which presumably find it inconvenient to raise and lower their opercula on a regular basis while living in cramped conditions. In tuna the opercula are rigid, because tuna do not pump water across the gills but ram it across while they are constantly swimming.

Two other external structures that deserve mention are breeding tubercles and cirri, even though they are found in comparatively small numbers of fish. **Breeding tubercles** are small, hard projections that appear on the head and fins of some fishes, especially minnows and suckers, when they are breeding. They are best developed in males and seem to be important as contact organs while the fish are spawning or as weapons used in defense of breeding territories. In eastern streams in the spring, it is common to find males of minnows such as stoneroller or hornyhead chub that have lost an eye as the result of a territorial dispute. **Cirri** are projections found mainly in the head region of many small, bottom-dwelling marine fishes. They often look like algae or attached invertebrates, hiding the fish from both prey and predators by making it look like a well-encrusted rock.

COLOR PATTERNS

Anyone browsing through a book featuring color plates of fish cannot help but be impressed with the enormous variety of colors and color patterns found on fish. Many of the patterns presumably evolved as ways for individuals of one species to recognize each other quickly, but most have much more meaning than that. What follows is a short catalog of some of the more common color patterns and their significance to fish.

Cryptic coloration. Many fishes are colored to match their backgrounds and hence approach invisibility. This cryptic coloration is particularly important for sluggish bottom fishes. As a result, they often match very closely the bottoms they rest on, even mimicking irregular patches of light or algae. Fishes with cirri use them to make their disguise three-dimensional. Many cryptically colored species are capable of changing their color to match their background. Some species of tide pool fishes may be green when living among green algae fronds but change to red when they live among red algae. The real masters of change, however, are flounders, which can quickly change their color pattern to match the bottom they are resting on. In the laboratory, some individuals have even produced a fair approximation of a checkerboard when placed in an aquarium with that pattern for the bottom.

Silvery. Bright silvery fishes are most characteristic of well-lighted waters, such as inshore waters of lakes or surface waters of the ocean. Usually silvery fishes school and the light flashing off the scales helps to confuse predators because individual fish become hard to pick out. The silvery color also helps the fish blend in with the light reflected from the surface of the water. The importance of silvery coloration is indicated by the fact that many stream fishes, such as trout and salmon, change from the complex color patterns they possess while in streams to a silvery color when they move into lakes or the ocean.

Countershading. Most fishes, but especially species that live in the water column, have dark backs and white bellies. The dark back helps them to blend in with the dark bottom or depths when viewed from above, whereas the white belly helps them to blend with the sky when viewed from below. The effectiveness of this countershading in providing basic protection from predators can be demonstrated by turning a hooked fish upside down in a lake or stream and noticing how conspicuous it becomes.

Disruptive coloration. Another form of camouflage is colors and patterns that break up the outline of fish, making them harder to see. One

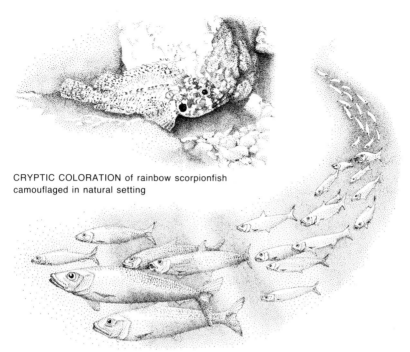

CRYPTIC COLORATION of rainbow scorpionfish camouflaged in natural setting

SILVERY APPEARANCE of school of Pacific sardine tends to confuse predators

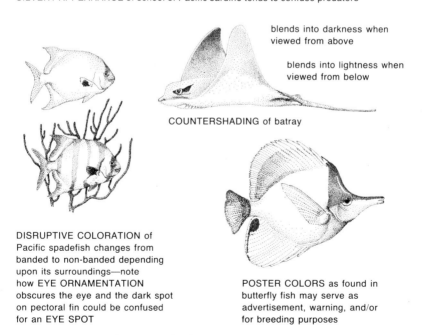

blends into darkness when viewed from above

blends into lightness when viewed from below

COUNTERSHADING of batray

DISRUPTIVE COLORATION of Pacific spadefish changes from banded to non-banded depending upon its surroundings—note how EYE ORNAMENTATION obscures the eye and the dark spot on pectoral fin could be confused for an EYE SPOT

POSTER COLORS as found in butterfly fish may serve as advertisement, warning, and/or for breeding purposes

Figure 2-7. Some functions of color patterns in fishes.

of the most common patterns of this type is vertical bars running down each side; this pattern is associated with fishes that live near beds of aquatic plants, such as bluegill sunfish or some cichlids. The vertical bars on the fish can blend in with the vertical pattern of the plant stems. Part of the camouflaging effect of vertical bars lies in the phenomenon of "flicker fusion." This occurs in the eye of the predator when a barred fish swims across a field of plant stems; the rapid movement of the bars causes them to appear to fuse with the plants and thus confuses the predator.

Red coloration.   In a world filled with predators, it seems strange at first that so many fish are bright red or have red spots, stripes, or fins. "Red snapper" (any of a number of species) is commonly sold in supermarkets; cardinal fishes and red squirrel fishes are among the most abundant fishes on coral reefs; the upstream migration of sockeye salmon, with their green heads and bright red bodies, is a frequently photographed spectacle. Although red is one of the most visible colors (to us) on land, it is frequently one of the least visible in water. Water is a selective filter of the colors of the spectrum, and red is filtered out in the first few meters. Red is also the first color to fade out at dusk and the last to appear at dawn. Thus most red fishes in the ocean live below the depths of red light penetration or are active only at night. Either way, they are difficult to see.

In shallow water, many fishes, such as sticklebacks and trout, use red as a breeding color. The fact that most such fishes confine the red to the males and develop it only during the breeding season is a good indication that it does make them more vulnerable to predators. This is balanced by the fact that bright color seems to make males more attractive to females. Red for such fishes represents a compromise between these conflicting demands of attracting females and avoiding predators because it is highly visible at close range or in bright light, but considerably less so at a distance (when viewed at an angle through water), or when the fish is under cover or in the shade.

Poster colors.   This term was applied by animal behaviorist, Konrad Lorenz, to the bright, complex color patterns characteristic of many coral reef fishes. Lorenz thought that the main function of these colors was to advertise the territories of the owners. We now realize that these colors have a number of possible functions, which are not necessarily exclusive of one another. In some fishes, such as butterfly fishes or puffer fishes, they may signal to predators that the possessor is too spiny or poisonous to be worth eating. The bright colors may also be important signs of sex, status, or maturity, especially in fishes that change their colors readily, such as the parrotfishes and wrasses. Curiously, another function of poster colors may be camouflage. Coral reefs are in fact dappled with patches of sunlight and encrusted with brightly colored corals and sponges with which the bright colors of the fish may blend when they seek the shelter of the reef. This is especially true when light levels are low and colors are hard to see; then the striking patterns may suddenly become disruptive patterns, making the fish hard to distinguish from shadow.

Eye ornamentation.   The eyes of a fish are perhaps its most visible feature, especially at a distance. They are frequently the focus of attacks by predators and are important in communication with other members of the species. This results in two contrary trends in eye ornamentation: one is to disguise the eyes, the other is to emphasize them. There are many ways of disguising eyes: running a black line through the pupil that is continuous with either horizontal or vertical stripes on the body; minimizing contrast between the iris and the pupil by having the iris dark in color, as well as the area surrounding the eye; having numerous spots surrounding the eye that are similar in size to the pupil. There are also many ways to emphasize the eyes. Most common is simply to have the eyes brightly tinged with blue, green, or yellow. Supplementary patterns, such as eye rings, are also common, at least in reef fishes.

Eye spots.   One of the most common marks on fishes, especially juvenile fishes, is a black spot located near the base of the tail. This spot is usually about the size of the eye and may even be emphasized with a light-colored ring, while the real eye is disguised. Their principal function seems to be to confuse predators, by having them aim for the tail rather than the head, giving the victim a greater chance to get away. In at least one species of South American cichlid, it has been shown that the eye spots of juveniles are the only thing that keeps the parents from eating them. So effective is this inhibitory effect that some species of minnows have adopted a similar eye spot to inhibit the predatory fish from eating them as well.

Lateral bands.   Single dark bands running along the side of a fish are best developed in schooling fishes. Their exact function is not known

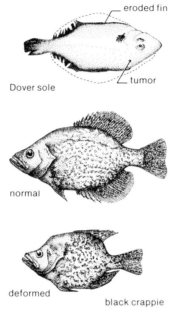

Figure 2-8. Abnormalities on fishes such as eroded fins, tumors, and deformed spinal columns are most common in fish populations subject to heavy pollution and other disturbances.

but they seem to help to keep members of a school properly oriented to one another. They also may help to confuse predators because the bands in a school may all blend together, making it difficult for the predator to pick out one individual.

ABNORMALITIES

Anyone who does a lot of fishing knows that it is not unusual to catch a fish that has some conspicuous abnormality such as a missing fin, a large sore on the side, or an odd-shaped head. In fact, the French taxonomist who first described the smallmouth bass gave it the Latin name *Micropterus*, meaning small fin, because the specimen upon which he based his description had an extra dorsal fin! Many of these abnormalities are natural, the result of accidents or developmental problems.

Usually abnormal fish do not last long because their different appearance makes them an easy mark for predators.

Abnormal fish are much more common today than a century ago and a high incidence of abnormalities may indicate severe environmental problems. Thus eroded fins and tumors are common on flatfishes caught near sewage outfalls in the ocean, because of their continuous contact with harmful materials on the bottom. An unusually high incidence of

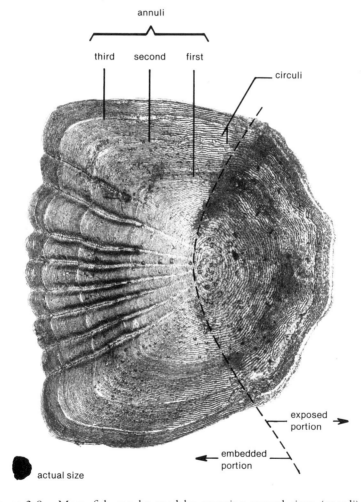

Figure 2-9. Many fish can be aged by counting annual rings (annuli) on scales. The annual rings are produced when steady growth, which produces the regular circuli, slows down or stops in response to low temperatures or other conditions. This fish was a little over three years old. (Photo by S. Woo, University of California, Davis.)

fishes with twisted backbones may indicate high levels of pesticides or other chemicals in the water. Such chemicals can affect developmental processes; they can also cause an early death to predators that concentrate the poisons contained in their prey. As a result, abnormal fish may grow up in the absence of predators and are not eliminated from the population. Abnormalities may also be the result of rearing fish in a fish hatchery because there are no predators except humans to pick out the abnormal fish that always appear in any batch of embryos. In addition, trout caught in a stream or lake that have abraded fins and sides probably were recently planted by the local fish and game department; their bodies reflect their constant contact with the sides and bottom of a cement rearing trough.

PROJECTS

1. Make a collection of fish from one body of water, using hook-and-line, a small seine, or other **legal** means, and examine the external features of all the fishes collected. On the basis of these features alone construct a diagram or chart showing where each species is likely to be found and what it is likely to feed upon. Check your results against a local guide to the fishes.

2. Find a place to watch fish with a mask and snorkel. Observe how they blend (or do not blend) into the background. Are small fish less conspicuously colored than large fish? What happens when a fish moves into a shadowed area? Fish often "hang out" underneath an anchored boat or in the shadow of a tree. Find out why by first observing the fish from a distance and seeing how close you can get before they flee. Then look outward from the shadowed area. Can you see further under water than before?

3. If you have access to a microscope, examine some fish scales under low power to see if you can tell how old the fish is by counting annual rings (annuli). The annuli are usually present as irregular bands or disruptions in the patterns of the regular bands (called circuli). If you compare the ages of fish of the same size and species taken from different environments, you can determine which environment is best for that species. Rapid growth reflects good conditions, so the fish of a given size should be younger than fish from less productive environments.

CHAPTER III

Fish from the Inside

As this book was being readied to go to press, three colleagues of mine at the University of California, Davis, created a minor stir in the scientific world by filming with a video camera the inside of a fish's mouth as it was feeding. Doctors Laurie Sanderson, Joseph Cech, and Mark Patterson inserted a tiny laser optic tube into the head of a Sacramento blackfish in order to discover how this fish manages to filter tiny particles from the water. Unfortunately, viewing the insides of a healthy, living fish is possible only with the aid of such extraordinary (and very expensive) technology. This means that to learn about internal anatomy of fish, you have to dissect them.

Fortunately, dissection of fish is extremely easy and many kinds of whole fish are often available at local fish markets. It is worth doing because the internal anatomy of a fish can tell just as much, if not more, about how it makes a living as the external anatomy. This chapter will therefore be a brief guide to the major internal organs of fish, with discussions of the structure and function of the major internal systems: digestive, circulatory, reproductive, respiratory, muscular, and skeletal.

MAJOR ORGANS

The major organs of fish are shown in the diagrams of a shark and of a bony fish (figures 3-2 and 3-3). Note the digestive tract that runs from the mouth to the anus, as the two variations shown here have a number

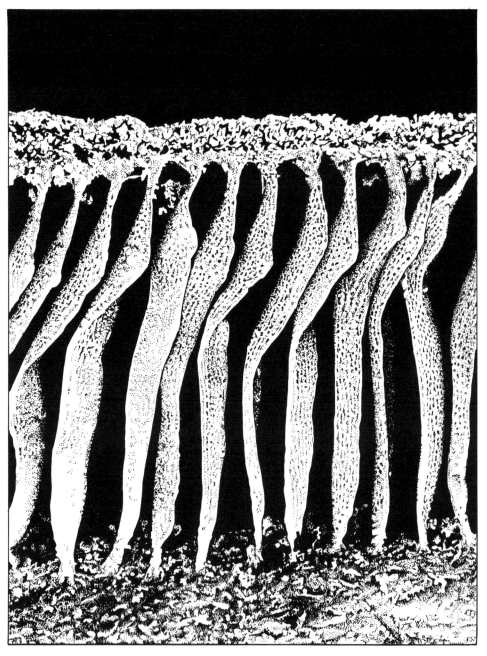

Figure 3-1.   Drawing of scanning electronmicrograph of the gill lamellae of a Sacramento blackfish.

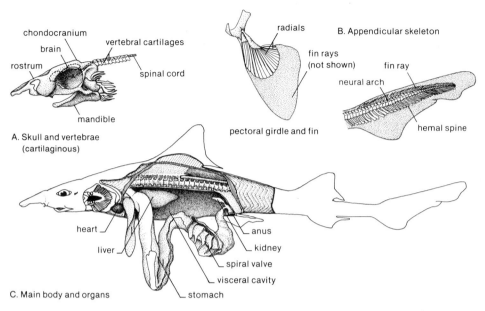

Figure 3-2.   Internal anatomy of a typical shark, the brown smoothhound.

of optional features not present in all fishes. Many fishes lack a true stomach, for example, although a stomach-like widening is usually present at the anterior part of the intestine. Some fish, such as trout, have **pyloric cecae** associated with the stomach; these are small, finger-like projections that may occur in some numbers. Note that the shark has a **spiral valve intestine**, which is a radically different design from the simple tube possessed by bony fishes. The spiral valve has a core around which the intestine walls curve like a spiral staircase. From the outside, the spiral valve looks like a second stomach. Associated with the digestive tract is what is usually the largest organ in the body, the **liver**. It is particularly large in some sharks, because it is used as a storage organ for lightweight oil that helps keep sharks buoyant. During the breeding season, the liver of bony fishes may be dwarfed in size by the gonads: the smooth white **testes** of the males, and the usually orange or yellow granular **ovaries** of the female. When fish are not breeding, the gonads are greatly reduced in size and often hard to find.

In most bony fishes, there is a large gas bladder that abuts the dorsal surface of the body cavity. In many fish it is thin walled and easily punctured, and so may not be obvious when the fish is cut open. In others, however, it may be thick walled or lined with muscles. Above the swim-

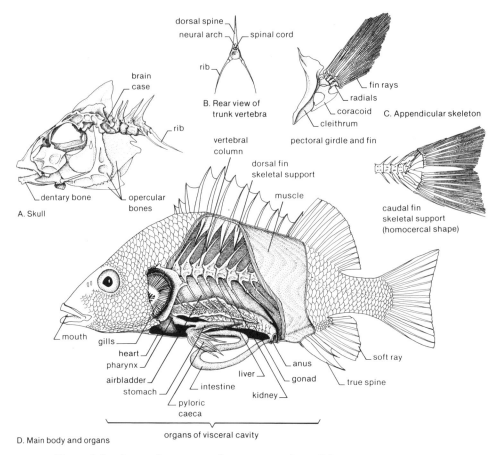

Figure 3-3.   Internal anatomy of a snapper, a bony fish.

bladder and running right along the backbone is the dark kidney; this is the dark material that is always so hard to remove when cleaning a fish.

Anterior to the main body viscera and more or less lying between the gills is the **heart**. The heart actually has four chambers:

1.   The thin-walled **sinus venosus**, which collects blood from the body before it enters the heart.

2.   The **atrium**, which receives blood from the sinus venosus and regulates its entry into the ventricle.

3.   The **ventricle**, which is the main pump of the heart.

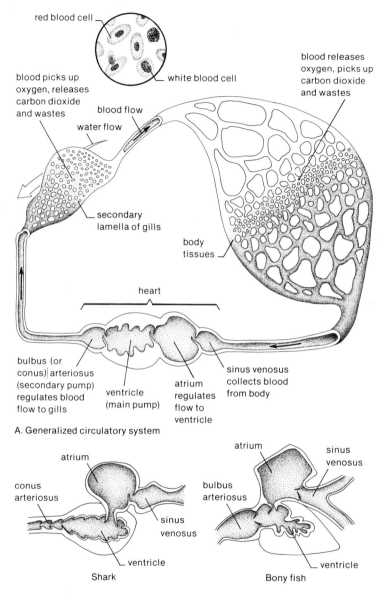

red blood cell

blood picks up
oxygen, releases
carbon dioxide
and wastes

white blood cell

blood releases
oxygen, picks up
carbon dioxide
and wastes

blood flow

water flow

secondary
lamella of gills

body
tissues

heart

bulbus (or
conus) arteriosus
(secondary pump)
regulates blood
flow to gills

ventricle
(main pump)

atrium
regulates
flow to
ventricle

sinus venosus
collects blood
from body

A. Generalized circulatory system

atrium

conus
arteriosus

sinus
venosus

ventricle

Shark

atrium

bulbus
arteriosus

sinus
venosus

ventricle

Bony fish

B. Cross-section of heart

Figure 3-4. *A:* Generalized diagram of blood circulation in a fish; the blood flows from the heart, through the gills, and then to the body. *B:* Cross-sections of hearts of shark and bony fish. Note the presence of two "extra" chambers in each heart: the sinus venosis, which collects the blood from the veins before it enters the atrium, and the bulbus arteriosus (bony fish) and conus arteriosus (shark), which function to help the muscular ventricle keep blood flowing through the gills at constant pressure.

4. The **bulbus** or **conus arteriosus**, a small supplementary pump that helps to keep the blood flowing to the gills at even pressure.

The gills, of course, are the most exposed portion of the internal anatomy because they are the principal site of gas exchange. Each gill consists of a series of gill arches that have gill filaments on one side and gill rakers on the other. The **gill filaments** are the soft material used in respiration; the **gill rakers** are hard and of various shapes and sizes, because they act like bars of a cage in retaining live food in the mouth cavity. In fishes that strain plankton and other small bits of material from the water, the gill rakers are long, slender, and packed close together. In fishes that feed on large prey, the gill rakers are short and stout. The size and spacing of the gill rakers in general reflect fairly closely the size of prey that each species favors.

DIGESTIVE SYSTEM

The digestive system consists of the gut and its associated organs, particularly the liver and pancreas. Food enters the system through the mouth and then passes through the straight **pharynx**, which in a few fish, such as herring and their relatives, may be modified into a muscular grinding chamber, much like the gizzard of birds. Much more commonly, the pharynx contains a set of powerful **pharyngeal teeth** that break up or grind food as it passes through. Pharyngeal teeth tend to be knife-like in fishes that are carnivorous, for shredding their prey. In fish that feed on plants or hard-shelled prey, such as common carp, the teeth are stout, with a striking resemblance to molars of mammals. In fact, pharyngeal teeth come in an amazing variety of shapes and sizes, reflecting the specialized diets of their owners.

The pharynx leads into the **stomach** (or its equivalent) in which the first stages of digestion take place. Large, well-developed stomachs are most typical of carnivorous fishes that capture large prey. From the stomach the food moves into the **intestine** where most of the breakdown of food into compounds that can be absorbed takes place. The length of the intestine, like the structure of gill rakers and pharyngeal teeth, reflects the diet of the fish. Long convoluted intestines, such as are found in suckers, carp, and mullet, are characteristic of species that feed on small, hard-to-digest items such as algae, organic matter, or small invertebrates. In contrast, most predatory fish have short intestines with only two or three bends in them.

CIRCULATORY SYSTEM

Compared to the digestive system, the circulatory system shows little variation in fishes. The heart consists of three or four chambers as described previously, which pump the blood through the gills, where it picks up oxygen and releases carbon dioxide (and other wastes). From the gills, the blood moves through a large artery (**dorsal aorta**) to the body, where it eventually is forced through capillaries before collecting in the venous system for return to the heart. The blood of fishes is similar to that of other vertebrates in containing both red and white blood cells. The red blood cells transport oxygen to the cells of the body by attaching oxygen molecules to the pigment hemoglobin. Fish physiologists have discovered that the capacity of fish hemoglobin to carry oxygen varies tremendously among species. Fishes with the most efficient hemoglobin, such as carp and Sacramento blackfish, usually live in warm waters with low oxygen levels.

REPRODUCTIVE SYSTEM

The gonads of fishes are paired structures that are suspended from the roof of the body cavity by membranous tissue. During the spawning season, the smooth white testes may make up over 12 percent of the weight of a fish, whereas the coarse yellow ovaries may make up 30 to 70 percent of the weight. The gonads of most fishes are usually conspicuous for three to six months of the year, so it is often possible to sex a fish by internal examination. The testes are rather similar in structure from one species to the next, but the way the sperm exit from the body shows more variation. In salmon and trout, for example, the sperm are simply shed into the body cavity and exit through a small opening near the vent. In sharks, the sperm enters a duct that is shared with the kidney and is often stored in a small chamber (**seminal vesicle**) for a while before being expelled. In most bony fishes, each testis has a separate duct to the outside. Eggs are passed from the ovaries in much the same manner as sperm is from the testes.

In species that have internal fertilization and subsequently bear young rather than laying eggs, the oviducts are highly modified. Examples of live bearers are many sharks, guppies, mosquitofish, and surfperch. When such fishes are nearly ready to give birth, the young may occupy an amazing amount of room and seriously impair the ability of the female to swim. In some live bearers, such as spiny dogfish shark and rockfish (*Sebastes*), the female simply incubates embryos that have their own food supply (yolk), whereas in others, such as hammer-head shark and surfperches, fairly elaborate connections exist between the embryo and the mother, much like the placentas of mammals.

RESPIRATORY SYSTEM

Gills are the principal site of gas exchange in fishes, but some fishes, such as eels, are capable of taking up oxygen through the skin. Still others can breathe air through lungs or similar organs. Respiration through gills presents some very interesting challenges to fish, because not only is the oxygen content of water much lower than that of air but the water must flow steadily across the gills in a direction opposite that of the flow of blood. The latter is necessary so the water will always be flowing across gill tissue containing blood that is lowest in oxygen. Despite these problems gills manage to extract up to 80 percent of the oxygen from water that flows across them. This is possible in part because gill filaments fit together like a sieve, assuring that a maximum surface area is exposed to the passing water. Flow across the filaments is kept constant by the fact that most fishes use two pumps in alternation to keep the water flowing: the opercula (gill covers) and the mouth cavity (with its fan-like branchiostegal rays). Some fast-swimming fishes, such as mackerel sharks and tuna, dispense with using pumps and simply ram the water across the gills. Consequently, if they stop swimming, they suffocate.

A number of fishes that live in stagnant waters breathe air. Lungfishes and gars actually have lungs for this purpose, but a number of other systems have evolved as well. The famous walking catfish of Asia (now of Florida, too) has its gills modified partially into an air-breathing or-gan that hangs from the roof of its mouth like an upside-down tree. The climbing perch of Asia and the longjaw mudsucker (a goby) of the west coast of North America have a well-developed blood supply to the roof of the mouth, which permits air breathing. Finally, a few South Amer-ican catfishes simply swallow bubbles of air and "digest" them in special regions of their guts!

MUSCULAR SYSTEM

In a vast majority of fishes, most of the body mass is made up of the large muscles of the body and tail that are used for swimming. Fishes

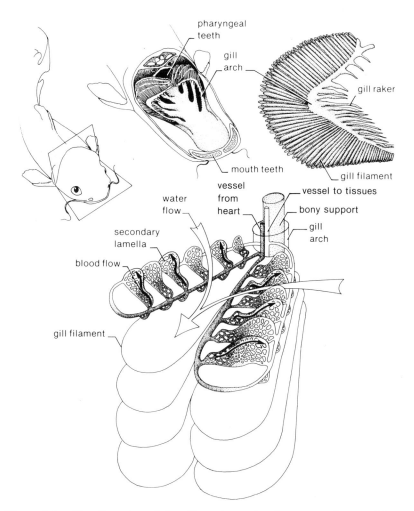

Figure 3-5. *Top:* Location of the gills and associated structures in a catfish. *Bottom:* Diagram of a section of gill, showing how water flowing in one direction passes blood (in the lamellae) flowing in the opposite direction, allowing oxygen to move from a fluid where it is at high concentration (water) to one where its concentration is low (blood).

that spend their time continuously swimming usually have a higher percentage of their body mass in these muscles than those that do not. Not surprisingly, these are the fishes generally most prized as food, for example, tuna, trout, bluefish, and striped bass. In contrast, fishes that swim largely by paddling with their fins have large muscles associated with the fins. Typical examples are the chunky puffer and porcupine

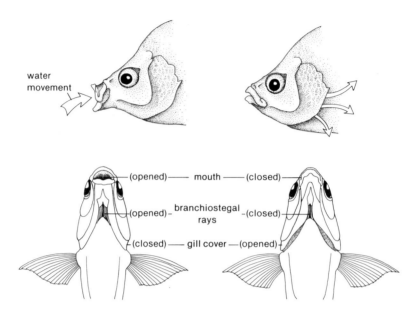

Figure 3-6.   Water flows across the gills of fish by means of a two pump system, which keeps the flow constant. The first pump is the mouth (buccal) cavity, which expands when the mouth is opened and water rushes in, and contracts when the mouth is closed, forcing the water across the gills. The second pump is the opercular cavity, which is created when the operculum (gill cover) is closed, causing water to rush into it, across the gills. These two pumps work together to assure that water is flowing across the gills at a steady rate.

fishes whose prickly defenses allow them to paddle about with a serene indifference to predators.

Another way in which the activity patterns of fish are reflected in the muscles is in the relative proportions of red muscle and white muscle. White muscle usually makes up the majority of the muscle mass and is often favored by humans because it does not taste as "fishy" as red muscle. The reason for this is that red muscle contains large amounts of blood, whereas white muscle does not. Red muscle is consequently what the fish uses for cruising and other constant or regular activities. The white muscle comes into action when a sudden burst of speed is needed, obtaining its energy from stored body sugars; these can quickly become exhausted, as when a fish hooked by an angler becomes "played out." Alternatively, a fish with a high (10–20) percentage of red muscle in its body (tuna, bluefish, striped bass) will spend more of its time swimming at high speed and probably put up a longer fight for the angler.

SKELETAL SYSTEM

The skeletal system of fishes has three major components: the vertebral column (backbone), the skull, and the appendicular skeleton of the fins and tail. The **vertebral column** can be solid cartilage, as in sharks, solid bone, as in most bony fishes, or a mixture of both. It provides the structural base for the muscles used in swimming, so there is generally one vertebra for each body segment or myomere. The trunk vertebrae in fishes are much simpler than those of terrestrial vertebrates because they do not need the various interlocking projections to counteract the pull of gravity. Many of the vertebrae of fish are nevertheless highly specialized. The two anteriormost vertebrae are modified to articulate with the skull, whereas the vertebrae on the other end of the spinal column are flattened and modified to articulate with the rays of the caudal fin. The trunk vertebrae in between have ventral ribs that extend downward between the muscle masses. Most fishes also have dorsal ribs that extend upward from the vertebrae. When such bones are well developed, as in suckers, carp, and pike, they can make the fish difficult to fillet and eat.

The **skull** of fishes is quite elaborate, because the head has many functions including:

1. Capturing and processing food.
2. Bringing in water for respiration.
3. The location of major sense organs.
4. A protective case for the brain and gills.
5. An attachment site of many muscles.
6. A streamlined prow necessary for efficient swimming.

In bony fishes, the result is a complicated puzzle of forty to sixty bones that show considerable variation in shape, size, and position from species to species. In sharks and their relatives the situation seems much simpler, because the braincase is just one solid piece of cartilage, molded around the brain and sense organs. The structure of this cartilage, however, shows considerable variation as well.

One of the most variable series of bones in the skull is that which makes up the upper jaw, reflecting the evolutionary shift in bony fishes away from a firm biting mouth lined with teeth toward a flexible mouth adapted for sucking in small prey. In the sucking mouth, the outer bones essentially function as lips, whereas the teeth are concentrated on more interior bones. Some fishes have no teeth in the mouth at all and rely entirely on pharyngeal teeth for processing food.

The **appendicular skeleton** is fairly simple; it consists of the internal supports for the fins, including the girdles underlying the pelvic and pectoral fins. The most elaborate series is that associated with the **pectoral girdle,** which starts as a series of five (usually) finger-like bones that support the fin rays and ends in a large bone (the cleithrum) that is united to the body muscles and joined by another bone to the head. The **pelvic girdle** is similar but less elaborate. The skeletal supports for the dorsal and anal fins consist of a series of small bones that lie between the muscle segments, one bone for each spine or ray.

PROJECTS

1. Carefully dissect a whole fish caught on a fishing trip or purchased in a market. Identify the major organs. Note especially the structure of the gill rakers and the length of the intestine. Cut open the stomach; are the food items present those that you would have predicted from the fish's anatomy? This exercise is particularly instructive if you can compare several species that are closely related, such as largemouth bass, bluegill sunfish, and green sunfish.

2. Next time you fillet a fish note the amount and location of its red muscles. How can you explain why there is such a small amount compared to the amount of white muscle?

3. Search a beach for fish bones and see if you can figure out what they are.

CHAPTER IV

Behavior

The dance . . . was continual. In from the sea, with an
emphatic marine delivery, the alewives put on their nearly
blind ceremony; and while I watched . . . everything seemed
attendant on them. The circle of the pond, and of the sea,
and of the needs of life, all the rounds of energy, in other
words, could center for this time on the cold-blooded fish.
There was an imperative rhythm in their spawning act, with
grace in its preparation and power in its fulfillment.
Humanity calls it love.

—*John Hay*, The Run*

Few things that fish do are as fascinating to people as the mass spawning runs of fishes such as alewife and salmon. In many areas people line the banks of coastal streams or crowd special viewing platforms to see spectacular events like those described by John Hay. We marvel at the internal urges that cause fish to enter the treacherous waters of a stream, often migrating hundreds of miles, blindly focused on reproduction. We wonder why so many of these magnificent fish have to die in order to reproduce. Like John Hay, we try to understand this behavior in terms of our own behavior.

Much of natural history is the study of behavior. In the case of fish, we either watch them directly to see what they are doing, or infer behavior from such things as body shape, stomach contents, or distribution patterns. Because most of the chapters that follow deal with how fish interact with one another and with their environment in a fairly narrow framework (e.g., trout streams, ponds), the purpose of this chapter is to present a broader picture of some of the more important aspects of fish behavior: migration, schooling, breeding behavior, aggressive behavior, feeding behavior, resting behavior, and communication.

*Published in 1965 (New York: Doubleday).

Figure 4-1. Karok Indian fishing for salmon at a falls on the Salmon River in California, using a large dip net. Elaborate Native American cultures developed on the west coast of North America because the migratory salmon were a large and predictable source of food.

MIGRATION

The predictable arrival of enormous numbers of fish on spawning or feeding grounds has made fish a dependable source of food for humans since prehistoric times. Native Americans of the Northwest Coast of North America built up dense populations and complex cultures based on the regular arrival of salmon in their rivers. At Clear Lake, California, the Pomo Indians knew when hitch and splittail made spawning migrations into the streams feeding the lake. They also knew when and where blackfish and thicktail chub spawned in the lake, so, like the Northwest Coast tribes, they too had a steady source of food. Similarly, Native Americans in the Midwest took advantage of runs of suckers and sturgeon, whereas those on the East Coast harvested abundant salmon, alewife, and shad that filled their streams in the spring. Early settlers also depended on these concentrations of fish, but it took only a few centuries for the runs to be depleted or destroyed by overfishing, dams that blocked access to spawning grounds, or destruction of spawning habitats and nursery areas.

Only recently has there been an interest in protecting the remnants of these runs and restoring some of the others. It is worth doing so because there are few sights in nature more spectacular than a run of fish, leaping with great purpose over barriers, and splashing, chasing, and courting on the spawning grounds, with gulls, osprey, and eagles circling overhead to feast on the dead or unlucky. Aesthetic values aside, these fish represent an enormous economic resource that provides returns year after year, with minimal investment except sensible management of our waterways, management that provides benefits far beyond the maintenance of spawning runs of fish.

Why do fish migrate? This is an important question because so many species devote enormous energy to migration, especially spawning migrations. Some fishes, such as Pacific salmon and American eels, devote so much of their bodily resources to migration and spawning that they die soon after. Eels cease feeding when they leave their freshwater homes for the sea. Once at sea they migrate thousands of miles, still without feeding, to spawn in the Sargasso Sea, where they die. Other species, such as steelhead trout and suckers, survive spawning but their energy reserves are so depleted they are much more vulnerable to predators, environmental stress, and disease. Such sacrifices are necessary to assure that the eggs are laid in an environment that maximizes survival of the young. Often an environment that is good for the adults is poor for the young, in part because the adults may eat their own young if they are readily available. Thus young salmon live in small streams that are rich in the tiny insects they need for growth. Similarly, the larvae (early life stages) of eels, sardines, and other fishes swim about in the rich soup of plankton they need for survival.

Because different life history stages of a fish have different food requirements, some fishes make a number of migrations in their lifetime, although some of the migration may simply be passive dispersal by currents. North Sea herring, for example, migrate to spawn in shallow areas with rocky bottoms off the coast of England. The newly hatched larvae drift across the English Channel to Germany and Denmark, where they congregate in the rich shallow coastal areas. As they increase in size, the "whitebait" move north. Finally, as juveniles, they swim out into the North Sea to join the adults.

Not all fish migrations are related to spawning. Albacore, for example, have a very narrow range of temperature preferences and follow the 14° C isotherm of the Pacific Ocean north in the summer and south in the winter. Temperature is also an important cue for herring and cod that move north toward arctic regions in summer to take advantage of the rich supply of food. By monitoring temperature, anglers and commercial fishermen can often predict with considerable accuracy when their favorite fishes will appear on the scene.

Temperature is only one of the cues that fish use to navigate when making migrations; they can also orient to gradients in salinity or other subtle changes in the chemical content of the water. For long-distance migration, fishes may use a number of methods. Some species may orient to the position of the sun during the day or to the direction of polarized light at dawn or dusk. Others may navigate in deep water by being aware of the earth's geomagnetic fields. For fish like salmon, the orientation mechanism does not have to be very precise, as long as it gets them swimming in the right direction. Once they reach a coastline they can depend on geographic features as well as temperature and chemical cues to help them find their spawning grounds. In the case of salmon, once they reach the region of their home river the distinctive odor of the stream where it mixes with ocean water brings the fish in; after they move upstream they locate the tributary stream in which they were spawned by its odor as well, and by its physical characteristics. These characteristics are "learned" (imprinted) by the young salmon as they migrate out to sea. The precision of these mechanisms for finding the traditional spawning grounds is truly remarkable, but a small per-

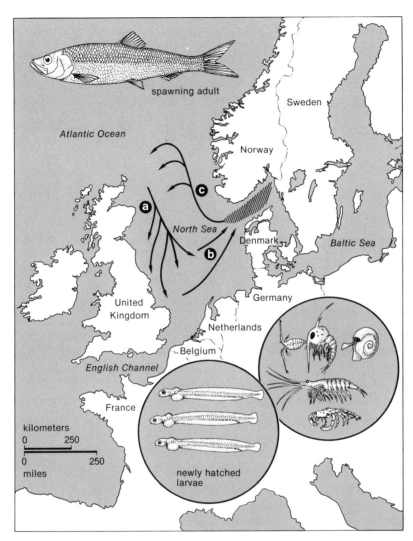

Figure 4-2.   Herring in the North Sea have regular migrations that separate the different life history stages. The adults (*a*) migrate from the North Sea to spawn off the coast of England. The newly hatched larvae (*b*) drift across to the bays and estuaries of the continent. Juvenile fish (*c*) eventually join adults in the North Sea, where they feed on abundant zooplankton (*circle on right*).

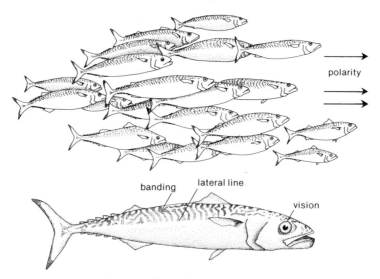

polarity

banding   lateral line

vision

Figure 4-3. Mackerels exhibit characteristics typical of schooling fishes: good vision, distinctive markings on the sides, and strong polarity of the schools.

centage of the salmon do make mistakes and wind up in the wrong streams. This is fortunate because it allows new streams to be colonized and old streams from which salmon have been wiped out to be recolonized.

SCHOOLING

Anyone who has kept tetras, zebrafish, or other small fishes in an aquarium knows that schooling is one of the more prevalent forms of fish behavior. Schools are groups of fish that are attracted to one another and are usually "polarized" in one direction. Individuals within each school are evenly spaced and a school may wheel and turn as if it were an organism itself. Schools can range in size from a handful of individuals to millions of fish. Schools of herring have been observed that occupied nearly 4.6 billion cubic meters of ocean. Naturally such schools are of considerable interest to humans, because fish are much easier to catch if they are concentrated. Indeed, the most important commercial fish species are schooling species.

Schooling depends to a large extent on vision, with each member of a school following some key feature of the fish around it, often a lateral stripe. Because of this dependence on vision, schools typically break up or at least lose their internal structure at night. Other senses, especially the lateral line system, also play a role in schooling behavior.

Fish school for a variety of reasons:

1. To reduce the risk of predation.
2. To increase the efficiency of food finding.
3. To increase reproductive success.
4. To increase the efficiency of swimming.

It might seem that schooling would actually make fish more vulnerable to predators rather than less, because a large school is much easier to spot from a distance than an individual fish. However, once a predator spots a school, it has to pick an individual fish out of the school, and that is very difficult. The positions of individuals change constantly as the school moves about and the light reflecting from the silvery scales makes one individual blend into another. In addition, once the predator approaches, hundreds of eyes in the school will be tracking it, so a surprise attack is nearly impossible. If the predator charges, the school will break fluidly around it as individual fish dodge out of the way and disappear into the mass of the school. The predator will be lucky if it does succeed in capturing a member of the school and will probably have expended an excessive amount of energy in the process, after repeated charges.

Not surprisingly, predatory fishes have developed many ways of overcoming the schooling defense. Sawfish and swordfish may simply charge into a school, thrash back and forth with their long bills, and then turn around to devour the individuals they have injured. Other predators, such as jacks, charge the school of prey in a school of their own. As the result of these simultaneous attacks, the school may break up into smaller schools and individuals become separated from the crowd, making them easy targets for the predators. Sometimes a school of predatory fish will trap a school of prey against the surface and many of the prey will leap out of the water in desperation, attracting terns, pelicans, and other predatory birds. The diving of the birds confuses the schooling fish even further and may allow the predators to feed to satiation. Swarms of diving birds are a common sight in coastal areas and on large lakes; anglers know that if they get to the feeding area quickly they may have excellent fishing for the predatory fishes! In tropical areas, flying fish presumably developed their extraordinary jumping abilities as a way of avoiding predators coming from below. This doesn't protect them from seabirds, of course, or from landing on the decks of passing ships.

Predators are also able to attack schools successfully by waiting until evening or early morning. When light levels are low, a predator can come closer to a school before being seen and is likely to find the school much less organized, with individuals easier to pick out. Thus it is common to find predatory fishes following a school of prey, even when they are not feeding. These fishes may, however, pick off individuals that lag behind because they are sick or injured.

Another advantage that schooling predators have is that when hunting a school of prey they can spread out and be much more likely to spot one. This same advantage applies to the small plankton-feeding fish they prey upon, because plankton usually occurs in patches and the patches are more likely to be found by a school than by an individual. For plankton feeders, it pays to be at the front of the school when a patch of food is located, because little may be left for those at the rear. Biologists observing schools of menhaden from the air have noted that they can be so efficient at filtering out the plankton from the water that the water behind the school is visibly clearer than the water in front of it.

Most fishes are not continuous schoolers like herring and menhaden, but many school around spawning time. Presumably, schooling for a period allows the fishes to synchronize their reproductive cycles, especially for fishes that spawn simultaneously in large numbers.

The hydrodynamic function of schooling is somewhat debatable, but the idea is that by spacing themselves properly within a school, each individual gets a boost from the swirl of water created by the fish in front of it. This theory is appealing, but careful measurements of the distances between individuals in schools do not always support it.

BREEDING BEHAVIOR

The breeding habits of fishes are as varied as the fishes themselves, so an account like this can only give you a glimpse of the variety. Eugene K. Balon of the University of Guelph in Canada has developed an ecological classification of breeding behaviors. According to Balon, there are five basic types of fishes, defined in relation to how they care for their eggs and young:

1. Scatterers
2. Brood hiders
3. Guarders
4. External bearers
5. Internal bearers

Scatterers are fishes that do not protect their eggs and young after spawning, but leave them scattered over the bottom, on aquatic plants, or drifting in the water. Most of these fishes are mass spawners, so the details of spawning behavior are difficult to observe: just a large swirling mass of fish. This is especially true of pelagic spawners, such as tunas, sardines, and anchovies. For freshwater fishes that scatter their eggs on the bottom, such as suckers and many minnows, the groups of spawning fish may be smaller; often the spawning act involves one female that is closely pursued by two or more males. Although most scatterers seem to have fairly simple spawning behavior, some of the most complex mating systems known involve this group, such as those of parrotfishes and wrasses of coral reefs. These systems involve such things as sex change, breeding territories, harems, and special males known as "sneakers" and "streakers."

Brood hiders also have no parental care of their young but they do make some efforts to protect their newly spawned eggs by hiding them, usually by burying them. The best-known examples of brood hiders are salmon and trout, which dig depressions (known as redds) in gravel riffles in which the eggs are laid, fertilized, and then buried. The females typically do most of the digging, whereas the males spend most of their time defending the redds against other males. After spawning is completed the fish either die (Pacific salmon) or leave (trout). Many minnows in the streams of eastern North America are brood hiders as well, although some, such as the hornyhead chub, build up piles of stones for hiding the eggs, rather than digging nests. A number of brood hiders do not do any construction but take advantage of ready-made situations. For example, a number of marine snailfishes lay their eggs in the gill cavities of crabs where they are both protected by the crab (to the crab's detriment) and provided with a steady source of oxygen. In a similar vein, the European bitterling (a cyprinid) deposits its eggs, by means of a long tube, in the gills of freshwater clams. In this case, the relationship is more reciprocal because the clams may simultaneously release their own larvae, which parasitize the gills of the bitterling!

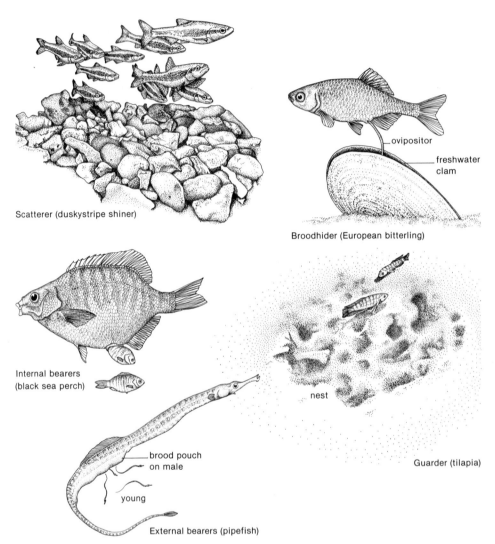

Scatterer (duskystripe shiner)

ovipositor

freshwater clam

Broodhider (European bitterling)

Internal bearers (black sea perch)

nest

Guarder (tilapia)

brood pouch on male

young

External bearers (pipefish)

Figure 4-4. Some breeding strategies in fishes. Duskystripe shiners spawn in schools, scattering their eggs in the gravel. Bitterling deposit their eggs on the gills of clams. Tilapia guard their eggs initially in nests and then may brood them in their mouths as well. Pipefish males carry the embryos and young in a special pouch. Black seaperch are true livebearers, nurturing the young with a placental-like arrangement.

Guarders are fish that protect their eggs and young after spawning, often investing a great deal of time and energy in the effort. Typically they produce fewer eggs than scatterers or brood hiders, but because of the parental care, the survival rate of the young is much higher. Guarding a brood usually requires guarding a spawning site as well, so most guarders are strongly territorial and have males that practice elaborate courtship rituals to lure the females to the spawning site. Unlike most terrestrial vertebrates, it is the male fishes that devote their time to parental care, not only guarding the eggs and young against predators, but keeping them free from debris and often keeping a current of oxygen-rich water moving over them as well. The main exceptions to the male parental care rule occur in the cichlids, in which both sexes may be involved in guard duty. Guarding may take only a few days or may last up to four months, as in the Antarctic plunderfish.

Guarders fall into two main categories, those that do not build nests and those that do. The dividing line between these categories is faint, however, because a number of fishes in the first category (e.g., gobies, lumpsuckers) will clean off the surface of a rock for the eggs and defend it. The best examples of guarders that do not build nests on the bottom are found among freshwater tropical fishes that live in stagnant water. The spraying characins, for example, spawn gymnastically on the underside of leaves that hang over the water. The eggs adhere to the leaves and the males stay underneath them, splashing water on them to keep them moist for the three days it takes the eggs to hatch. Such eggs not only have an excellent supply of oxygen, but they are relatively free from predators. The Asiatic gouramis, such as Siamese fighting fish, paradise fish, and climbing perch, guard their clusters of embryos in the oxygen-rich layer that exists at the surface of the water. The embryos themselves each contain a lightweight oil droplet and so float at the surface after spawning. In some species, the male provides them with additional protection by surrounding them with oxygen-rich bubbles which it blows. This is necessary because the waters in which the fish live are often stagnant and the adults themselves breathe air.

Nest builders alter the substrate in one way or another to provide a place to shelter the eggs and young. In North America, the most visible of the nest builders are the sunfishes and black basses (family Centrarchidae) which dig saucer-shaped depressions in shallow water, often in large colonies. The males defend their nesting territories vigorously against other males and predators; in the colonial nesting species, such as bluegill, the best sites are in the middle of the colony where egg predators are fewest. The females patrol the colonies and seem to prefer to spawn with the males with the most vigorous displays, the strongest color patterns, and the best nest sites. Studies by Dr. Mart Gross show that this leads to cheating on the part of some small males that otherwise would get little opportunity to spawn. These males hang out on the fringes of the nesting colony and dash in when they observe a pair of fish spawning, quickly mingling some of their own sperm into the spawn. However, the chewed-up fins and poor condition of many of the "sneaker" males indicate the high cost of this reproductive strategy. Still other small males spawn by adopting female color patterns and behavior, which allow them to swim slowly into nests and join spawning pairs by convincing the dominant males that the intruder is female.

Other nests are constructed of loosely woven pieces of aquatic plants, such as those made by sticklebacks. The male threespine stickleback, in particular, constructs its nest and courts females so readily in captivity that stickleback behavior has become among the most studied of all vertebrates. Briefly, the male makes a loose tunnel of plant material glued together with a special secretion from his kidney. Once the nest is finished, he attempts to attract the attention of passing females by the bright red color of his breast and his ritualized courtship behavior. Each female is lured into the nest by a special zig-zag "dance"; there she lays a few eggs. The male then chases the female out of the nest and fertilizes the eggs. After spawning with several females, each male proceeds to incubate the eggs by fanning them with his fins. After the young hatch and leave the nest, the male guards them for a while, until they become too independent to "herd."

Although the behavior of exposed nest builders is best studied, the most common types of nest are those in caves, cavities, or burrows. Fishes living in the intertidal zone of the ocean, such as pricklebacks and gunnels, will guard eggs in caves underneath rocks, wrapping their bodies around the eggs to keep them moist if the tide drops too low. In streams, sculpins (Cottidae) and darters (Percidae) will stick their eggs to the underside of flat rocks in midstream where there is enough space to guard them and enough current to keep them well oxygenated. In ponds, fathead minnows will spawn underneath old logs and boards; each male develops a special pad on top of its head which is used for rubbing algae and debris off the undersurface on which the eggs are deposited. Catfishes, such as channel catfish, will occupy hollow logs, old muskrat burrows, and other cavities for the guarding of their embryos. After the young hatch, the school of young will be guarded for a week or so; the swirling balls of young are common sites in catfish spawning areas, with the guarding adult close by.

External bearers are fishes that carry the embryos about after spawning. Perhaps the most common types are mouth brooders, found in a variety of species, from sea catfishes to cichlids. Mouth brooders carry the large, yolky embryos in their mouths and, in cichlids, have the young use the mouth as a temporary refuge as well. A less common type of external bearing is found in pipefishes and seahorses, in which the eggs are carried about on the bellies of males. In the most primitive arrangement, found on some pipefishes, the eggs are simply stuck to the belly and carried about until they hatch. Other pipefishes have grooves or open pouches to hold the eggs and young. Male seahorses (which are closely related to pipefishes) have a closed pouch with only one opening in which the eggs and young are kept until they can fend for themselves.

Internal bearers carry the embryos and/or young inside the female and have internal fertilization as well. Some internal bearers simply deposit the developing embryos in the environment shortly after fertilization. This is characteristic only of a few obscure groups of bony fishes. Some sharks and many rays retain the embryos for a number of weeks after fertilization but eventually deposit them in the outside world wrapped in tough, leathery cases with tendrils on each corner for attaching to seaweeds. When they break loose they may wash up on beaches as "mermaid's purses." The next advancement in internal bearing is **ovoviviparity**, in which the embryos are retained by the female until they are ready to swim on their own. The female provides no extra nutrition for the developing embryos beyond that which is present in the yolk sac of the eggs. Examples of ovoviviparity are found in many large sharks, the coelacanth, and marine rockfishes (*Sebastes*).

The most advanced form of livebearing is **viviparity**, where the developing young are provided with extra nutrients by the mother, often through a placenta-like arrangement. The most mammal-like of these placental arrangements is found in the requiem sharks (Carcharinidae) and in the hammerhead sharks (Sphyrnidae). Other sharks, such as the sand tiger, have a more direct way of providing nutrition to the developing young: the first young to hatch in each of the two uteri devour the rest of the embryos present and subsequently feast on unfertilized eggs provided by the mother. In the marine surfperch family (Embiotocidae), the embryos receive nutrition from the mother through their enlarged fins, which are in intimate contact with the ovarian wall. In guppies and mosquitofish (Poeciliidae), it is pericardial (heart-lining) tissue

Figure 4-5.   A stiff, arched body, flared gill flaps, and extended fins are good indicators of aggressive behavior in fishes, as shown here in tule perch.

of the embryos that develops nutritional contact with the ovarian wall of the mother.

The males of internal bearers have either the anal fin (bony fishes) or pelvic fins (sharks, skates, and rays) modified to permit internal fertilization of the females. Courtship is usually brief, although male guppies are renowned for the frequency with which they attempt to copulate with females.

AGGRESSIVE BEHAVIOR

One of the most frequently observed aspects of reproductive behavior is aggression, especially where one male defends its territory against another male, or chases another male away from a receptive female. Much of such behavior is highly ritualized and is accompanied by modified swimming, flaring of gill covers and fins, and even altered color patterns. The ritualization reduces the likelihood of injury to the fish involved, unless the fish are especially bred to fight (as in Siamese fighting fish) or unless there is no place for the loser of an encounter to go (as in small aquaria).

Aggressive behavior is also frequently observed in relation to feeding.

One of the consequences of introducing fish food at the same place in an aquarium is sometimes to have one of the larger fishes defend the area as a feeding territory. This can also be observed in young salmon and trout in streams, which may defend vigorously the best places to feed, usually near both cover and fast water. It has been observed in such territorial fish that larger fish always dominate smaller fish, but that if two fish are about the same size, the fish first in the territory usually dominates. Species may also make a difference; in eastern North America, brown trout have replaced brook trout in many streams, because they are more aggressive. They are able to keep brook trout out of the best hiding spots, making them much more vulnerable to anglers and other predators.

FEEDING BEHAVIOR

The feeding behavior of fishes is determined in large part by their morphology. A fish with a downward pointing mouth will feed largely on the bottom, whereas a fish with a terminal mouth is more likely to pursue prey in open water. When a particularly rich source of food is abundant, however, a fish may switch from its normal mode of feeding to take advantage of it. In lakes, fishes with many different body shapes can be observed feeding on the surface when a large hatch of insects is in progress. I have even observed suckers, fish very well designed for staying on the bottom and sucking up algae and small invertebrates, skimming the surface of a lake in order to suck up pine pollen floating on the surface. Trout, which normally feed in the water column, may switch to foraging on the bottom if some insect is abundant or if the water is too turbid to see drifting prey.

Recently, it has become fashionable in scientific circles to discuss feeding behavior in terms of **optimal foraging theory**. This theory states that fish will forage in such a way as to maximize the amount of energy they take in while minimizing the effort to do so. The less energy spent on feeding, the more there is available for growth and reproduction. Feeding, by this theory, can be divided up into components that each consume energy:

1. The amount of energy spent searching for prey.
2. The amount of energy spent pursuing prey.
3. The amount of energy spent handling captured prey.
4. The amount of energy spent digesting the food.

An optimal foraging strategy is one that balances the energy expended in each area so the total amount of energy spent is minimized. Thus, it may be beneficial for large trout to feed on small insects only if those insects are so abundant that they can be captured without much energy being spent on search and pursuit; otherwise it pays the trout to stay quiet and wait for larger prey to come by. Largemouth bass may choose to feed on slender, soft-rayed minnows, even if less abundant, rather than on deep-bodied spiny-rayed sunfish because the handling and digestion time is likely to be higher for sunfish.

Of course, many other factors influence foraging behavior besides energy balance. One of the most important is to avoid being part of some other creature's optimal foraging. Dr. Mary Power observed that algae grew most thickly in shallow water in Panamanian streams because grazing catfish were likely to be eaten by birds if they ventured in there to feed; similarly, thick patches of algae were not consumed by stonerollers (an algae-eating minnow) in an Oklahoma stream if the patches were near deep water favored by predatory largemouth bass. Around coral reefs, there is often a ring of sand in which no turtle grass and algae can grow because of grazing by herbivorous fishes; a short distance from the reef the grass may grow in luxuriant meadows because cruising predators keep the herbivores out.

RESTING BEHAVIOR

The fact that fish spend a considerable amount of time "sleeping" is difficult for humans to accept because fish do not have eyelids and so cannot close their eyes. Yet most fish do spend a considerable amount of their time in an energy-saving, quiescent state. This became obvious to me only when I ventured into scuba diving at night in a Minnesota lake. Minnows, which I observed in large active schools during the day, at night were scattered over the bottom in shallow water, moving back and forth with the motion of waves. Sunfish were observed in loose aggregations, slowly circling. Bass and perch, their daytime colors washed out, were seen resting on or under logs. Even with a bright light, I could get quite close to these fishes and even touch them on occasion. A similar phenomenon can be observed on coral reefs. The brightly colored, day-active fishes move into shelter as darkness sets in, where they remain quietly for the night. Parrotfishes may even secrete a blanket of mucus about themselves to make it more difficult for night-hunting eels to find them. On reefs, a whole set of night-active fishes, such as squirrel and soldier fishes, stay quietly in the crevices of the reef during the day, behaving much like day-active fishes do at night.

Fishes such as tuna and mackerel sharks, which must keep swimming in order to keep breathing, presumably must swim at good rates of speed all night long. Schooling fishes such as herring, however, swim about slowly in the water column and the daytime school becomes only a loose aggregation.

COMMUNICATION

It should be obvious from the previous discussions that visual cues are extremely important in fish behavior. This emphasis on vision is at least partly due to our own visual bias; other types of communication are much more difficult to observe. The importance of sound, for example, is just beginning to be recognized. We now know that many fishes have courtship songs produced by such means as vibrating the swim bladder with special muscles and rubbing bones against one another. Sea catfish make a loud low-frequency sound when they move their pectoral fin spines in their sockets, a sound that can be amplified by the swim bladder. Drums (Sciaenidae) get their name from the loud regular sounds they can produce by vibrating their swim bladders.

Fishes may also communicate by means of chemical odors. Yellow bullhead catfishes apparently can recognize each other as individuals by odor and associate these odors with position in the local social hierarchy. Cichlid fishes can recognize their own young by odor for at least three weeks while they are herding them around. In many species it is likely that the sexes are attracted to one another partly through the use of **pheromones**, special compounds evolved just for that purpose. Another class of compounds produced by some fishes is **fear scents**. These are released into the water when a predator kills or wounds an individual; when other members of the species detect a fear scent they immediately take evasive action. This mechanism is particularly useful in water where visibility is limited.

Still another method of communication among fishes is through the use of electrical signals. Each time a muscle contracts it gives off a small electrical charge. Because water is a good conductor of electricity, some fishes have developed certain muscles specifically for the production of electrical signals, which are often their principal means of communication. This is particularly true of the gymnotid fishes (including the electric eel) of South America and the mormyrid fishes of Africa. These fishes communicate with one another in turbid water by varying the pulse width and frequency of their electrical discharges. They have specific electrical signals for courtship, aggression, and individual recognition, using electrical signals much like birds use songs.

PROJECTS

1. Find out the location and timing of spawning runs of fishes in your area. Spend some time observing the behavior of the fish. Bridges are often good vantage points for such observing. You do not have to be near an ocean to see spawning runs, as many freshwater fishes also make spawning migrations.

2. If available, collect some sticklebacks and keep them in aquaria to observe their reproductive behavior, which is described and illustrated in many books on fishes. Another species that is readily available (in bait shops) and that breeds easily in captivity is the fathead minnow. Some species of darters (e.g., logperch) may also spawn in aquaria if conditions are right.

3. In an aquarium, introduce three sunfish or cichlids of different sizes and observe the dominance relationships. Do the same thing with three fish of the same size. In the latter case, see what happens when you allow one fish to become established before you introduce the others.

CHAPTER V

Diversity

Angel shark. Hagfish. Sarcastic fringehead. Warmouth. Whitefish. Grayling. Cardinal tetra. Wobbegong. Peacock flounder. Hogchoker. Zebrafish. The colorful names we give to fish reflect their enormous diversity. There are over 21,000 species of fish, with new species being described on a regular basis. They occur in an amazing array of habitats from high mountain streams to the depths of the ocean, with a diversity of adaptations to match their habitats. This chapter, and the ones that follow, can only give you a glimpse of this diversity, which is equal to that of all mammals, birds, reptiles, and amphibians combined.

CLASSIFYING FISH

We give fish names as a way of organizing our knowledge about them. With so many species, it is also necessary to further organize this knowledge by developing rather artificial classification systems for them. The purpose of the systems is also to show the relationships of different groups of fishes to one another. Because our knowledge of fish characteristics is constantly improving, the classification of fishes is constantly changing, although the underlying structure remains fairly constant. For example, the largemouth bass has the Latin species name *Micropterus salmoides*, a unique combination of names that allows scientists all over the world to know what species it is. *Micropterus* is the genus name and indicates that the largemouth bass is closely related to other "black" basses, such as the smallmouth bass (*Micropterus dolomieui*). It is

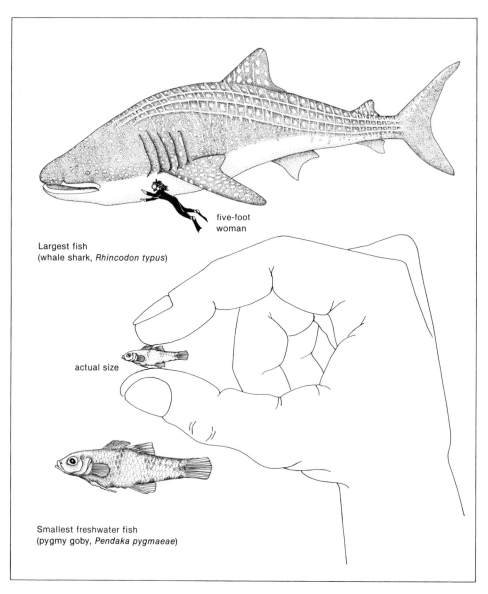

five-foot
woman

Largest fish
(whale shark, *Rhincodon typus*)

actual size

Smallest freshwater fish
(pygmy goby, *Pendaka pygmaeae*)

Figure 5-1. The largest fish (18 m long) and one of the smallest fishes (12 mm long when mature).

placed in the family Centrarchidae, which is made up of the black basses, sunfishes (members of the genus *Lepomis*), crappies (genus *Pomoxis*), and a variety of other species that share many characteristics and a common recent ancestry. At the next highest level of classification, the largemouth bass is in the order Perciformes, a large group of advanced perch-like fishes with spines in their fins. Going up still further, the Perciformes are part of the subclass Actinopterygii, the fishes with rayed (rather than lobed) fins, which is part of the class Osteichthyes, the bony fishes. Above this is the superclass Vertebrata, the phylum Chordata, and the kingdom Animalia. The complexity of classification systems is greatly increased by the use of further subdivisions, such as infraclass, superfamily, subfamily, and subspecies.

Scientific names were developed as a way of standardizing animal and plant names independent of common names, which vary from place to place. *Oncorhynchus tshawytscha* is chinook salmon, king salmon, or tyee, depending on where you are. Today in North America common names are often more constant than scientific names because the American Fisheries Society has produced a list of standard common names and these names have become widely accepted by scientists, anglers, and naturalists. Chinook salmon, for example, is the name now preferred to king salmon or tyee. The name rainbow trout has been standard for at least fifty years, whereas its scientific name has changed from *Salmo irideus* to *Salmo gairdneri* to *Oncorhynchus mykiss*! The last change took place in 1988 after scientists concluded that (1) rainbow trout are more closely related to Pacific salmon (*Oncorhynchus*) than they are to Atlantic salmon (*Salmo*) and (2) the rainbow trout of Asia are the same as those found in North America. Since the Asian trout were described first, their name (*mykiss*) took precedence, using the firm set of rules established by the International Commission on Zoological Nomenclature.

LAMPREYS AND HAGFISHES

Lampreys and hagfishes (class Agnatha) are not true fishes because they lack such amenities as paired fins and jaws. Nevertheless they are generally included with them because they are aquatic vertebrates that are frequently collected with (on or in) fishes. Although they are highly specialized, they also possess many features in common with the presumed ancestors of all vertebrates and so are very useful to students of vertebrate evolution. Both lampreys and hagfishes are long and eel-like and

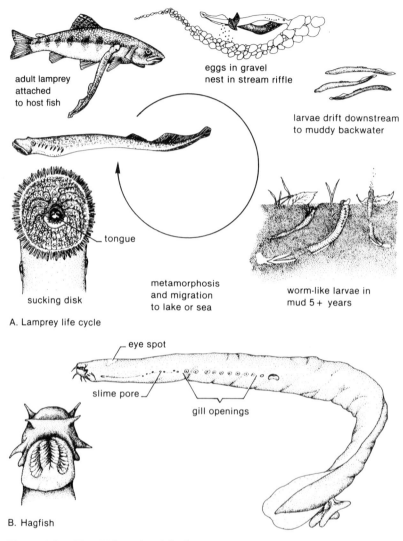

adult lamprey
attached
to host fish

eggs in gravel
nest in stream riffle

larvae drift downstream
to muddy backwater

tongue

sucking disk

metamorphosis
and migration
to lake or sea

worm-like larvae in
mud 5 + years

A. Lamprey life cycle

eye spot

slime pore

gill openings

B. Hagfish

Figure 5-2. *Top:* Life cycle of the lamprey. *Bottom:* A hagfish.

possess many primitive characteristics, but otherwise represent two to-
tally different evolutionary lines going back 400 million or more years.

Lampreys as adults are readily recognizable by the sucking disc on
the mouth that is covered with sharp "teeth," the line of gill slits along
each side of the head, and the large eyes. In the middle of the sucking
disc is a powerful tongue that ends in sharp teeth. Adult lampreys are

predators on the bony fishes that replaced their less specialized ancestors. A lamprey latches on to the side of its victim, holding on with its
sucking disc and rasping a hole with its tongue; it then sucks out blood
and other bodily fluids. Larger fish often survive lamprey attacks; in
areas where lampreys are common, it is not unusual to find fish with
healed lamprey scars. However, when sea lamprey from the east coast
of North America invaded the Great Lakes, following the construction
of the Welland Canal around Niagara Falls, they devastated the native
fish populations. These fishes were not adapted to the lamprey's particular style of predation. Only after millions of dollars were spent on research, on lamprey eradication programs, and on restocking programs
were fisheries restored to these lakes; these efforts are still ongoing.

Like salmon, adult lampreys migrate upstream where they construct
nests in gravel riffles, spawn, and die. The tiny young that emerge from
the nest are the behavioral opposites of their predatory parents. They
first drift downstream until they are washed into a muddy backwater.
They then burrow into the mud and spend five or more years living a
worm-like existence, sucking up algae and detritus from the surface of
the mud. They are black and eyeless, so the transformation to the large-
eyed, silvery adult form is dramatic.

Hagfish are also worm-like in appearance, with no obvious eyes. The
great taxonomist Linneaus, in fact, placed them in the phylum he had
erected for worms. Hagfishes are entirely marine. They burrow into the
bottom and feed mainly on clams, worms, and other invertebrates. They
are actually very abundant in many temperate areas and come to the
attention of commercial fishermen because they are scavengers as well
as predators. As scavengers, they will burrow into a dead or dying fish
and consume it from the inside out. Because they are unable to discrim-
inate between fish dead of natural causes and those caught in nets, they
are frequently inadvertently caught themselves, when fish they have burrowed into are hauled up on deck. As if to add insult to injury, not only
will the fisherman find that he has been robbed of his prey, but the
hagfish will have secreted large amounts of sticky slime on the deck.
The capacity of a single hagfish to produce slime is truly amazing. The
inside of their skin is lined with slime glands; the slime is excreted
through tiny pores. The natural function of the slime seems to be to
protect a source of food, such as a dead fish, from other scavengers,
which apparently do not like the slime either.

Today hagfish are actually in danger of being overfished in some parts of the world because there is a big demand for leather made from their tough skin. Usually the leather is sold at high prices as "eel skin." Presumably "hagfish skin" would have little market appeal.

SHARKS, SKATES, AND RAYS

The elasmobranch fishes (class Chondrichthyes) are much maligned and misunderstood, thanks to the general fascination with the results of encounters between swimmers and large sharks. Even though sharks do manage to kill a handful of people every year, the probability of getting mangled in an automobile accident on the way to the beach is much higher than the probability of being attacked by a shark, even in "shark-infested" waters. However, as L. J. V. Compagno, one of the foremost experts on sharks explains:

> [T]he world shark catch for 1976 was about 307,085 metric tons. If the average shark in this catch weighed approximately as much as the average human being—say 68 kilos or 150 pounds—the catch would be the equivalent to sharks "catching" 4.5 million people. Clearly, sharks have much more to fear from people than vice versa. (*Oceanus* 24 [1981]:6)

Elasmobranchs are also undeserving of the label "primitive" that frequently gets assigned to them. Instead, they are a group of fishes that has had a long and independent evolutionary history during which they developed different solutions to the problems of being a fish from those developed by bony fishes. These adaptations are covered in detail in a number of readily available books (see chapter 15), so will not be mentioned here, except for one example, the sensory systems. As anyone familiar with shark lore knows, elasmobranchs have an excellent sense of smell. Less appreciated is the fact that this sense is integrated with the other sensory systems in a way that makes sharks particularly effective predators. Thus they have a sense of hearing that is very sensitive to low-frequency vibrations, a well-developed lateral line system, and vision that is acute at low light levels and sensitive to color. They also have a special electrical sense that detects the minute electrical field that surrounds all living creatures. As a result, sharks will actually close their eyes, with a special membrane, just before striking their prey because they can sense the electrical fields, and so do not need to see their prey.

The elasmobranchs consist of about seven hundred species and exhibit a surprising diversity of body shapes and ways of making a living. The classic sharks, including those that attack humans, are mostly re-

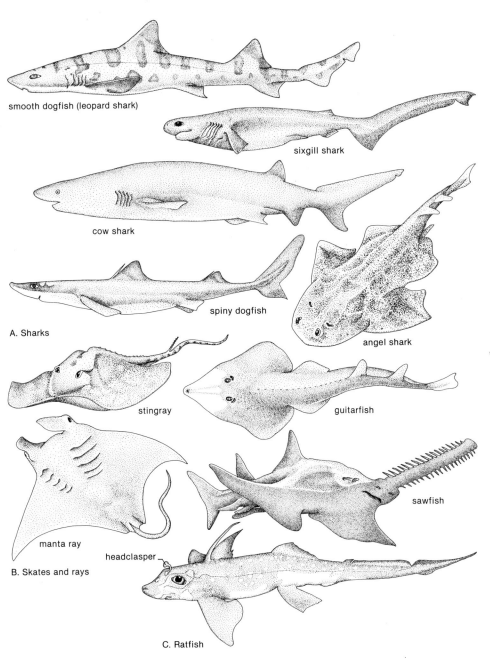

smooth dogfish (leopard shark)

sixgill shark

cow shark

spiny dogfish

angel shark

A. Sharks

stingray

guitarfish

manta ray

sawfish

headclasper

B. Skates and rays

C. Ratfish

Figure 5-3.   Chondrichthyes: Sharks, skates, rays, and ratfish.

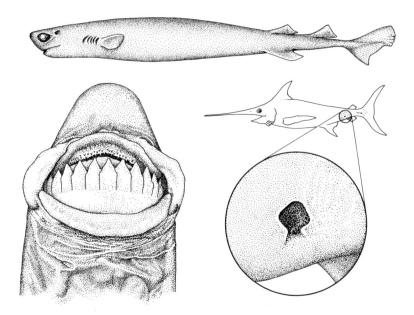

Figure 5-4.   The cookie-cutter shark is a small shark that ambushes large fish such as swordfish and bites out cookie-sized chunks of flesh with its special-ized teeth.

quiem sharks (family Carcharhinidae) and mackerel sharks (Lamnidae); these are large, wide-ranging predators on fish and marine mammals. Closely related to these sharks are the two largest fishes, the whale shark, which can attain lengths of 18 meters, and the basking shark, which reaches 14 meters. Both of these species swim about with their mouths open, filtering plankton through special strainers inside.

Although these large sharks form much of the public image of sharks, much more abundant are the many small species of sharks, such as the spiny dogfish (family Squalidae) which is commonly dissected in college anatomy laboratories. They are also increasingly common as the fish in fish and chips. The smallest known sharks are also in the Squalidae; one species can become mature when only 11 centimeters long. These small sharks are seldom encountered, however, because they occur in deep water. Some species are covered with photophores (light-producing or-gans) in patterns that mimic those of luminescent squid; this apparently allows them to get close enough to the squid to catch them. It may also allow them to turn the tables on large fish that dive into the depths to feed on squid, because several of these small shark species attack large fish, cutting round chunks of flesh out of their victims' sides with special teeth. These sharks have become known as cookie-cutter sharks as a result.

One of the more peculiar sharks is the angel shark (Squatinidae), which is flattened much like a skate or ray and lives on the bottom much as they do. However, it can be distinguished from skates and rays by the fact that its gill openings are on the side of the head, rather than on the bottom, and its wing-like pectoral fins are not continuous with the head. Skates and rays are superbly adapted for life on the bottom. They have two spiracles on top of their heads which draw in water that passes through their gills and is expelled through the ventral gill slits. Their fins are wing-like and can be used for flying through the water or for digging out buried clams and other prey. They rely mainly on camouflage for protection from predators, but stingrays (Dasyatidae) have a toxic spine toward the end of their whip-like tails which they can use effectively.

There have been many modifications on the basic skate and ray design. The eagle rays (Myliobatidae), for example, have developed strong muscular pectoral fins for cruising about. This is also true of the mantas (Mobulidae), the largest of all rays, which have a peculiar funnel-like mouth for feeding on plankton. The electric rays (Torpedinidae) have converted some of their muscles into electric organs, which they use to jolt their prey into submission. Large electric rays can produce up to 200 volts at a time. Another unusual means of capturing prey is that of the sawfish, a large ray that uses a long snout with a row of sharp teeth on either side to slash its way through schools of fish. It then goes back and eats the pieces and wounded fish. Curiously, this method of prey capture evolved independently in the saw sharks (Pristiophoridae), which are related to the dogfish sharks.

RATFISHES

The ratfishes or chimaeras are a group of about twenty-five species of Chondrichthyes that have a long and separate evolutionary history from sharks and rays, so are placed in their own suborder (Holocephali). They dwell on the bottom, mainly in fairly deep water, where they feed on clams and other invertebrates that they crush with their strong, flat teeth. The head and eyes are large, but the tail is long and slender.

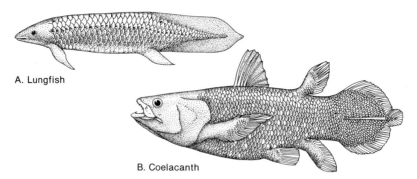

A. Lungfish

B. Coelacanth

Figure 5-5. Two modern lobe-finned fishes that give some idea of what the fish ancestors of amphibians looked like.

Among their curious anatomical features are the presence of gill covers (opercula), which are otherwise found only on bony fishes, and a clasper, used to hold the female during copulation, on the head of each male.

BONY FISHES

Most of the world's fishes are bony fishes (class Osteichthyes). They are so diverse in form that it is difficult to find characteristics that are common to all species. Basically they have a common structural pattern that includes the presence of lungs (or a swimbladder, derived from lungs), bone, bony scales, and segmented fin rays. The diversity of bony fishes can be appreciated by examining some of the many families within the two major evolutionary lines of bony fishes, the lobe-finned fishes (Sarcopterygii) and the ray-finned fishes (Actinopterygii).

LOBE-FINNED FISHES

The lobe-finned fishes exist today as only about seven species but they are of enormous interest because they seem to be descendents of the fishes that gave rise to terrestrial vertebrates. The principal feature that supports this idea is the limb-like fins, which have supporting bones that attach to the pelvic and pectoral girdles. Six of the seven lobe-finned fishes are lungfishes (order Dipneusti), which include one species in Australia, one in South America, and four in Africa. All breathe air through true lungs, which enables them to survive periods of stagnant water and drought. The African lungfishes actually burrow into the mud of a drying lake and secrete a cocoon of mucus, with a tiny opening to the surface. They then become extremely torpid and will survive in the co-coon for many months, until the next rain allows them to emerge.

The chief rival to the lungfishes in the tetrapod ancestor look-alike contest is the coelacanth (order Crossopterygii). This remarkable fish was first discovered in 1938, when one was caught by a fisherman off the east coast of Africa. It caused great excitement because coelacanths had been known previously only from fossils, the most recent being 70 million years old. Modern coelacanths are large (up to 2 meters) pred-ators on fish and squid that live on the bottom in deep water. The back of the head is hinged, giving the mouth extra gape for swallowing large prey. They presumably rest on the bottom on their lobed fins, to ambush their prey. Their evolutionary path has taken them so far from their inland origins that their lung-like swimbladders are filled with fat. The first coelacanths were freshwater fishes inhabiting tropical swamps and presumably breathed air. By the age of dinosaurs (Jurassic) they had largely shifted to being predators in shallow marine habitats. They dis-appeared from the fossil record along with the dinosaurs and were thought to be extinct until rediscovered, associated with deep water reefs.

RAY-FINNED FISHES

In terms of number of individuals and number of species the ray-finned fishes (subclass Actinopterygii) are the dominant fishes today, and have been for over 300 million years. These fishes have fins that are attached to the body by fin rays (rather than lobes), branchiostegal rays (part of their improved respiratory apparatus), and no internal nostrils (unlike lungfish). Except for about thirty-three species of "primitive" forms, all of this group are teleosts, a comparatively recently evolved group. Among other characteristics, the teleosts have a symmetrical (homocer-cal) tail, thin scales, a swimbladder (rather than lungs), highly maneu-verable fins, and jaws that are adapted as much for sucking as for biting. The remainder of this chapter will discuss important families of ray-finned fishes, to provide an idea of their diversity.

Sturgeons (Acipenseridae) are "primitive" bony fishes that have a car-tilaginous skeleton, lack scales except for a few that have been modified into rows of ridged bony plates along the back and sides, and have a shark-like heterocercal tail. There are only twenty-three species but many of them are (or were) quite abundant, moving between salt and fresh water or large lakes and streams and feeding on whatever they can

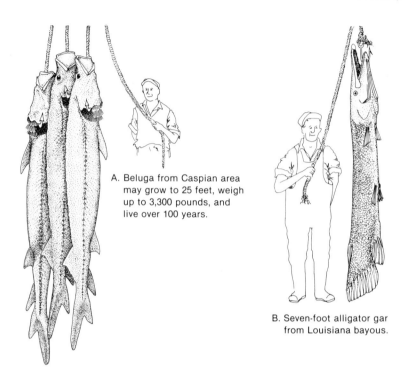

A. Beluga from Caspian area
may grow to 25 feet, weigh
up to 3,300 pounds, and
live over 100 years.

B. Seven-foot alligator gar
from Louisiana bayous.

Figure 5-6.   Sturgeons and gars are bony fishes of ancient lineage that seem
to have persisted through time by being large and specialized.

suck up off the bottom. One factor contributing to their success is the
enormous number of eggs produced by large females; this has also con-
tributed to their decline because the eggs are prized as caviar. They are
among the largest fishes occurring in fresh water. The beluga sturgeon
of Russia reaches 8.5 meters and 1,300 kilograms, whereas the white
sturgeon of western North America may reach 4 meters and 590 kilograms.

   Gars (Lepisosteidae) are another ancient group. They are confined to
North American fresh waters and are among the most distinctive of
freshwater fishes, as their cylindrical bodies are covered with an armor
of hard, diamond-shaped scales. Their bony heads are tipped with a
long snout full of sharp teeth for grabbing the fish they feed on. Gars
catch their prey from ambush; they are classic lie-in-wait predators. The
heavy armor that protects gars is possible because it is balanced with a
large air bladder that keeps them buoyant. The air bladder also functions as a lung, which is handy in the warm stagnant waters gars often

prefer. Most gars reach 1 to 2 meters in length, although the alligator gar may reach 3 meters.

HERRINGS, EELS, AND ELECTRIC FISHES

Herrings (Clupeidae) are teleosts that were once considered primitive but now are considered just to be highly specialized for schooling and plankton feeding in open water. This family contains about one hundred eighty species, including shad, sardines, gizzard shad, and menhaden, and is closely related to the anchovy family (Engraulidae). Among the adaptations of herrings are large silvery scales that reflect light to dazzle daytime predators. These scales are also deciduous, so if a predator does grab a fish, the victim may escape by leaving a few of its scales in the predator's mouth. The belly of herrings tapers to a sharp keel, so light streaming down from above does not create the faint belly shadow characteristic of more rounded fishes. This makes them less visible to predators attacking from below.

Eels (order Anguilliformes) are much more diverse than most people recognize; there are over six hundred species belonging to twenty-two families, occupying habitats from coral reefs to deep sea to fresh water. They all possess long tapering bodies, small, usually wedge-shaped, heads, and sharp teeth. Most are adapted for living in crevices or burrows, so lack features that might make them unable to wiggle out of a tight situation; they lack one or more sets of paired fins, loose scales (they are tiny and deeply imbedded), large opercular bones (they have adopted a less efficient way of pumping the water across the gills), and the flexible, lip-like bones of the jaws that characterize most teleosts. Moray eels (Muraenidae) are perhaps the most familiar eels because their toothy heads are frequently photographed peering out of holes in reefs, but the most studied eels are the freshwater eels (Anguillidae). Members of this family are highly prized food fishes over much of the temperate world and can be caught in large numbers when they migrate from fresh water to salt water for spawning. In fresh water, they are secretive predators on fishes and invertebrates, reaching 35 to 150 centimeters in six to twelve years. As maturity approaches, they transform from being a hard-to-see green color to being silvery, a color more advantageous in the open waters of the ocean. Both European and American eels (east coast) migrate thousands of kilometers in the ocean to

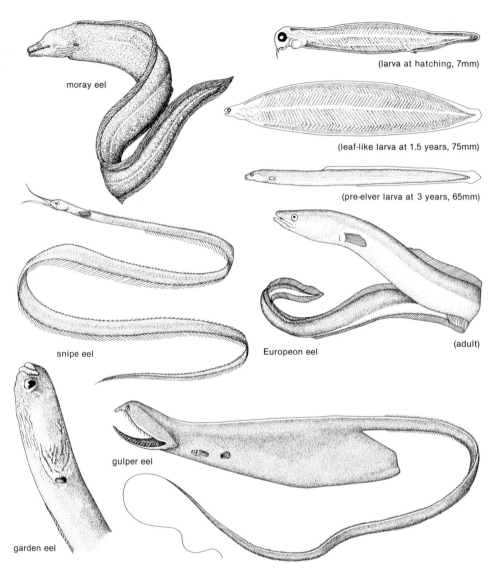

moray eel

(larva at hatching, 7mm)

(leaf-like larva at 1.5 years, 75mm)

(pre-elver larva at 3 years, 65mm)

snipe eel

Europeon eel

(adult)

gulper eel

garden eel

Figure 5-7.   The over six hundred species of eels show a surprising variety of forms and ways of making a living; all, however, pass through the larval stages shown in the upper right.

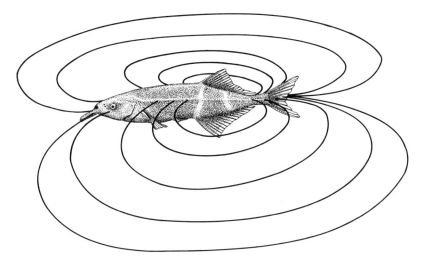

Figure 5-8.   The elephant fish is an African electric fish (family Mormyridae). It lives in muddy waters and is most active at night, finding its way about with electrical sense organs. It is sensitive to anything that enters the weak electrical field (indicated by lines around fish) produced by special electric organs.

the Sargasso Sea, where they spawn in deep water. The main evidence we have of this spawning is the appearance of small leaf-like larvae that drift in surface currents for one to three years until they reach the proper coastline. They then transform into elvers, some of which may migrate long distances upstream to a lake or pond, where they grow to adulthood.

African electric fishes (Mormyridae) are the most species-rich family of the ancient order Osteoglossiformes (bony-tongues). This odd order was once one of the more abundant groups of teleosts, but today exists as a few species on each continent. The only ones left in North America are the mooneye and goldeye (Hiodontidae), two herring look-alikes that are abundant in the upper Mississippi drainage. The African electric fishes, with over one hundred species, live in the muddy rivers and swamps of tropical Africa. In situations where good eyesight is of little use, they locate prey and communicate with one another using electrical signals. Curiously, in some species the brain-to-body ratio is about the same as that of humans.

PIKES, SALMON, AND BRISTLEMOUTHS

Pikes (Esocidae) are lie-in-wait predators distantly related to salmon and trout. They are found only in northern Eurasia and North America,

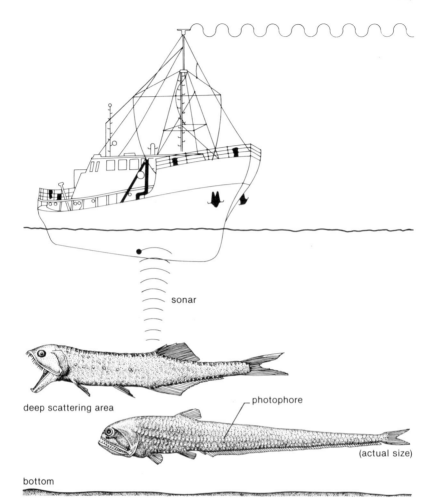

Figure 5-9. Bristlemouths are small deep-sea fishes that are probably the most numerous fishes in the world. At times they are so abundant that sonar from ships reflects from them, rather than the ocean bottom, giving a false reading of depth.

in slow-moving streams, weedy lakes, and castle moats. One species, the northern pike, is found on both continents and is an important food and sport fish wherever it is found. It occasionally reaches 1.3 meters in length, on a diet of fish. The largest of the pikes is the muskellunge of eastern North America, which reaches 1.6 meters (32 kilograms). Pike capture their prey with a sudden rush from cover, grabbing it sideways with their sharp teeth. They then turn the prey around and swallow it headfirst. Pike are most vulnerable to capture in the spring, when they move into shallow areas of flooded vegetation to spawn.

Salmon, trout, whitefish, and grayling (Salmonidae) make up a family of only about one hundred fifty species, but the literature on them is the most voluminous of any fish family. One reason for this is that they are often the dominant fishes in the streams and lakes of north temperate regions where most fish biologists reside, as well as being important oceanic predators in these areas. A more important reason, however, is that they are exceedingly valuable as sport and commercial fishes. Their significance to European and North American anglers is almost mystical and has led to such arcane activities as catch-and-release fly fishing and the planting of domesticated trout in suitable waters all over the world.

Salmonids are superbly adapted for living in cold water and many species are highly migratory, so they have been able to quickly colonize streams created by the rise of mountains and the melting of glaciers. Thus they occur in some of the most remote and spectacular regions of the northern hemisphere, contributing to their mystique. There is little need to say more about these fish, because examples of their behavior and ecology are scattered throughout this book.

Bristlemouths (Gonostomatidae) are fishes few people have seen, yet they may be the most numerous fishes in the world. They are tiny (less than 5 centimeters) deep-sea fishes and are so abundant in places that the sonar beams of submarines may reflect off their swimbladders, creating the impression that the bottom is much closer than it really is. This phenomenon is known as the deep scattering layer. Bristlemouths are distantly related to salmonids, having an adipose fin. They have large mouths with spiky teeth and fine gill rakers; this allows them to capture a wide size range of prey, although they feed mainly on zooplankton.

CHARACINS, MINNOWS, AND CATFISHES

Characins (Characidae) are a large family of freshwater tropical fishes that includes such popular aquarium fishes as the tetras, as well as the notorious piranhas. The roughly twelve hundred species in this and sixteen smaller related families are recognizable by their adipose fin and distinctive teeth in their jaws. They are representative of the superorder

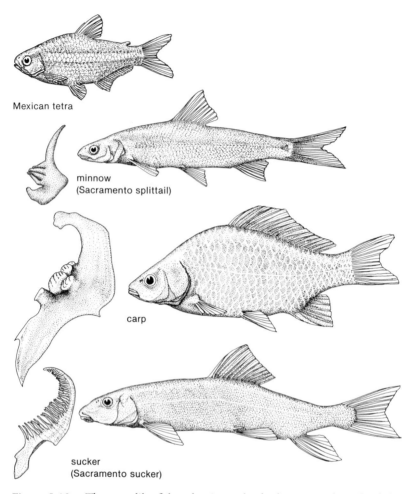

Mexican tetra

minnow
(Sacramento splittail)

carp

sucker
(Sacramento sucker)

Figure 5-10.   The carp-like fishes dominate the fresh waters of much of the
world. One reason for their success is the pharyngeal teeth, located in the
"throat," that permit highly specialized diets. *From top:* A small tetra, typical
of South American waters and a minnow with pointed pharyngeal teeth used
for chewing up insects and other invertebrates; the common carp, with its
molar-like teeth that are used for crushing a wide variety of food; a sucker,
with its comb-like teeth that are used for breaking up masses of algae and
small invertebrates.

Ostariophysi, which contains the fishes that dominate the fresh waters of most of the world, including the minnows (Cyprinidae), suckers (Catostomidae), loaches (Cobitidae), South American electric fishes (four families), and catfishes (thirty-one or more families). There are five to six thousand species in this superorder (including sixteen hundred cyprinids and two thousand catfishes), about one-quarter of all known fishes.

There are many reasons for the success of this group but foremost among them are their acute sense of hearing, their pharyngeal teeth, and their fear scent. **Acute hearing** is very useful in fresh water where visibility is often limited. In this group of fishes, it is made possible by the Weberian ossicles, a small chain of bones that connects the swimbladder (which vibrates with sound) to the inner ear. This is remarkably like the system that gives mammals such an acute sense of hearing, where a chain of bones connects the eardrum to the inner ear. **Pharyngeal teeth** are teeth located in the throat, behind the gills, that are used for grinding, chewing, or shredding food once it has been swallowed. This allows these fish to have delicate mouths specialized for prey capture, especially by sucking, placing the heavy teeth well back in the head. Just as mammals have achieved success by having teeth specialized for many different food types, the pharyngeal teeth of minnows and catfishes are specialized for many different food types. The **fear scent** is a chemical given off by an injured fish that immediately causes its fellows to take evasive action by diving for cover or making a tight school. This is a useful mechanism in turbid water, where predators are difficult to detect visually.

Minnows and carps (Cyprinidae) make up the largest single family of fishes, yet most of the sixteen hundred–plus species are rather generalized fish: fusiform bodies, large eyes, visible scales, pelvic and pectoral fins ventrally located and well separated from one another, and small mouths with bony, flexible "lips." Not surprisingly, these fishes are mainly day-active predators on small invertebrates, although a number of species are specialized for feeding on other fish or on algae and detritus. Most species are small (less than 10 centimeters) but some, like the Colorado squawfish, attain lengths of nearly 2 meters. They are typically the most abundant freshwater fishes in North America, Eurasia, and Africa, but are absent from South America (which is dominated by characins instead). In North America, there are about two hundred species of small cyprinids called shiners, many of them members of the genus *Notropis*. These fishes are silvery most of the year but may develop bright red breeding colors in the spring, when the males of many species build and defend nests. Carp and goldfish are two species of cyprinids that have been widely introduced around the world; part of the reason for their success is the presence of stout spines on the dorsal and anal fins, which makes them much harder for predators to consume.

Suckers (Catostomidae) are a small (about sixty species) but very successful family in North America, with two species in Asia as well. They are close relatives of the minnows and are specialized for sucking and scraping algae, small invertebrates, and organic debris from the bottom. To do this they have extendable, fleshy lips that point downward, comb-like pharyngeal teeth for breaking up mats of algae and debris, and long intestines for breaking down hard-to-digest material. Most reach 40 to 80 centimeters long as adults and make spawning migrations to favored riffle areas of streams, where they congregate and are vulnerable to predators (and humans). They are frequently accused of competing with trout because declines in trout are often associated with increases in suckers. Almost no good evidence exists that such competition occurs; instead the increase in suckers is caused by an increase in conditions favorable for them, such as warmer water and increased algal growth (usually as the result of human activity).

North American catfishes (Ictaluridae) are but one small (forty species) and relatively undistinguished family in the extraordinary worldwide array of catfishes. There are catfishes covered with bony plates of armor, catfishes that exceed 5 meters in length, catfishes that are so tiny they can slip into the gill cavities of larger fish and nibble on the gills, catfishes that breathe air, and catfishes that live in the ocean. The North American forms are plain by comparison but do possess the basic catfish features which have made the group so successful:

1.   Secretive night-oriented behavior.

2.   Long whiskers on the snout for finding prey and mates in the dark.

3.   Smooth, scaleless bodies.

4.   Stout spines leading the dorsal and pectoral fins, often with a venom gland at the base of each spine.

At night, catfishes rule the world that cyprinids and characins rule during the day. A number of these catfishes are favorite food and sport

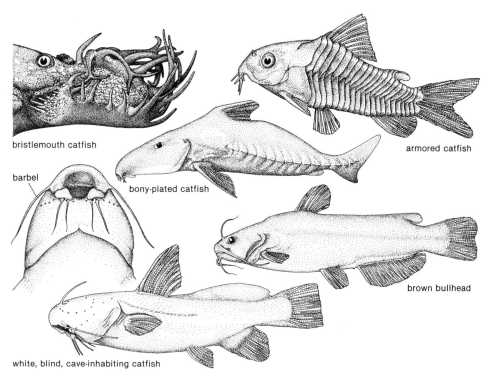

bristlemouth catfish

armored catfish

barbel

bony-plated catfish

brown bullhead

white, blind, cave-inhabiting catfish

Figure 5-11.   Catfishes come in an astonishing variety of shapes and sizes but most are active at night and find prey through the use of the sensitive, whisker-like barbels on their snouts.

fishes, especially channel, white, and blue catfishes and black, brown, and yellow bullheads, which thrive in warm turbid water. The madtoms are small catfishes common in eastern streams and are much more abundant than most people realize. They have given rise to a number of pale, blind catfishes that inhabit streams in caves.

LANTERN, COD, AND ANGLER FISHES

Lantern fishes (Myctophidae) are another important contributor to the deep-scattering layer of the deep ocean (see discussion under bristle-mouths). They represent a diverse order of deep-sea fishes (Myctophi-formes) that includes such curiosities as spiderfishes, paperbones, dag-gertooths, pearleyes, and telescope fishes. The lantern fishes are so named because they have rows of light organs (photophores) on their bodies and heads, which they use to recognize each other in the dark.

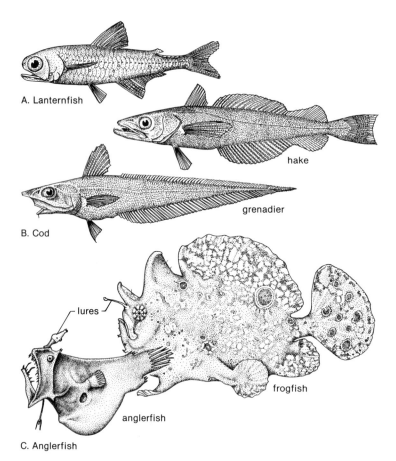

Figure 5-12. Representatives of some important marine orders. *A:* Lanternfish (order Myctophiformes), abundant deep-sea fishes and important food of tunas. *B:* Cods (Gadiformes), with the hake as a representative of a commercially important species and grenadier as a deep-sea representative. *C:* Anglerfishes (Lophiiformes), which use lures to ambush prey in a variety of habitats.

They are small fishes with blunt heads and large eyes. They make extensive vertical migrations every day in pursuit of the small invertebrates they feed upon.

Codfishes (Gadidae) are the most economically important members of the order Gadiformes, a diverse group of about seven hundred species of bottom fishes. The fifty-five species of cod, haddock, and pollock have long supported important fisheries, especially in the North Sea and off the coasts of North America. Codfishes are elongate fishes, with three dorsal fins and a chin barbel. Their large mouths and sharp teeth indi-

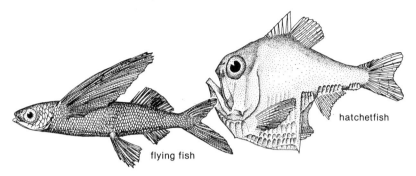

hatchetfish

flying fish

Figure 5-13. Two unusual marine fishes, the flying fish, which uses its long pectoral fins to increase the distances it can leap from the water, and the hatchetfish, a common deep-sea fish, whose hatchet-like profile makes it nearly invisible to predators from below. The fishes are not drawn to scale. Hatchetfish are typically around 8 cm long and flying fish are 25 or more cm.

cate their diet of other fish, crabs, and molluscs. Many species of codfishes make extensive seasonal migrations, in response to changing temperatures, salinities, and abundances of favorite prey. Thus, off Newfoundland (originally "found" by European fishermen as a place to dry and salt cod) the arrival of cod was often predicted by the arrival first of schools of capelin, a small fish important in the cod's diet.

Anglerfishes (order Lophiiformes) are odd distant relatives of the cods. They are lumpy, flattened, or globular fishes, each with a fishing pole (illicium) growing out of the head, tipped with a lure (esca). The esca can mimic anything from a small fish to a copepod and is waved about in front of the angler. Fish attracted by the lure are engulfed in the huge mouth. Most anglerfishes are small deep-sea fishes and are among the most bizarre of all fishes. Most are black, football-shaped fishes, with huge fanged mouths and tiny fins. One of the shallow water forms, however, is the goosefish, which weighs up to 32 kilograms. Despite its ugly appearance (by our standards), this fish is very tasty and is sold as monkfish, "the poor man's lobster."

Flying fishes (Exocoetidae) are plankton-feeding fishes that have evolved an excellent way to escape predatory fishes. They leap out of the water and glide as much as 100 meters with their wing-like pectoral fins. They can extend the glide by sculling the surface of the water with the lower lobe of the tail, which is considerably longer than the upper lobe. Thus flying fishes do not use their wings to power their flight but only for gliding. This is in contrast to the only real flying fishes, the fresh water hatchetfishes (Gasteropelecidae), which are related to the chara-

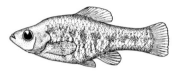

Figure 5-14. A pupfish, a type
of killifish confined to springs
and other aquatic habitats in
North American deserts.

cins. The name of these fishes comes from their hatchet-shaped breast,
which contains the large muscles needed to move their pectoral fins rapidly up and down, to achieve true flight.

Killifishes (Cyprinodontidae) are about three hundred species of
small fishes that live in some of the most inhospitable aquatic environments around the world, such as temporary ponds, salt marshes, desert
springs, salty lagoons, and even mountain lakes. They survive because
of their extraordinary tolerances for fluctuating temperatures and salinities, their ability to eat a wide variety of food, including bluegreen algae,
and their ability to reproduce at an early age. There are a number of
species that live in temporary ponds in Africa and South America; they
persist by depositing their eggs in the bottom mud of the pond. The eggs
in the mud resist desiccation even when the pond dries up, so that as
soon as the pond fills again, they hatch. They then grow rapidly; at least
one species can go from egg to spawning adult in four weeks!

In North America, the most familiar killifishes are the pupfishes (*Cy-
prinodon*), found in places like Death Valley, and the topminnows (*Fun-
dulus*), typical of eastern streams and salt marshes. A family that differs
from the killifishes mainly in their method of reproduction is the Poeciliidae (livebearers). This family includes such hardy aquarium fishes
as guppies, swordtails, and mollies, as well as mosquitofish. The latter
species, native to the southeastern United States, is now found worldwide because of its ability to survive and breed in the same habitats as
mosquitoes, which it devours (along with everything else that moves,
including the young of native fishes).

STICKLEBACKS AND SCORPIONFISHES

Sticklebacks (Gasterosteidae) are perhaps the most recognizably fish-
like members of an otherwise peculiar-looking order (Gasterostei-
formes) made up mostly of fishes covered with armored plates and hav-

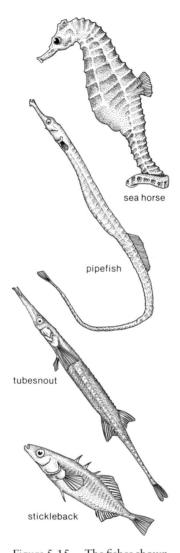

sea horse

pipefish

tubesnout

stickleback

Figure 5-15.   The fishes shown
here are all related to one another   (order   Gasterostei-
formes), although the seahorse
is one of the most unfish-like of
all fishes. This shows how the
seahorse presumably evolved
from a more conventional fish
ancestor, using modern-day
forms as models for the various
evolutionary stages. Both the
stickleback and the tubesnout
have a row of bony plates on
each side, whereas the pipefish
and seahorse are encased in
bony plates.

ing extremely elongate bodies, for example, tubesnouts, pipefishes, sea-horses, and trumpetfishes. Sticklebacks are small, mostly freshwater fishes with sharp spines on the back and fins and a row of bony plates on each side in lieu of scales. The evolution of stickleback-like fishes into seahorses can be imagined by first looking at tubesnouts, which look like elongate, armored sticklebacks, and then at pipefishes, which look like attenuated tubesnouts encased in wrap-around armor. Sea-horses, which are in the same family (Syngnathidae) as pipefishes, are essentially pipefishes whose bodies have assumed a permanent S curve, so they can swim upright, moving about by sculling with their dorsal fin. Seahorses, like pipefishes and sticklebacks, feed on tiny inverte-brates they suck up from aquatic plant beds and the water column.

Rock and scorpion fishes (Scorpaenidae) are the "typical" members of the thousand species order Scorpaeniformes,which also includes sea robins, sculpins (below), poachers, snailfishes, and lumpsuckers, all of which have a close attachment to the bottom. Like most members of the order, the rockfishes (*Sebastes*) have large heads and mouths, large round pectoral fins, rounded caudal fins, and extremely spiny dorsal and anal fins. The spines have venom glands associated with them that in rockfishes can make a puncture wound very sore but that in some trop-ical scorpionfishes (such as stonefish) can be lethal to humans. Rock-fishes are (or were until overfished) extremely abundant off both North American coasts, although the Atlantic has only two species compared with over sixty for the Pacific. Many of them are red in color, because they live at moderate depths, and are marketed as "red snapper."

Sculpins (Cottidae) are small, smooth, large-headed fishes found mainly in two environments: swift cold streams and the rocky intertidal zone. Both these environments are rich in food but very turbulent, so sculpins are adapted for living underneath rocks or clinging to the bot-tom while the currents sweep over them. Sculpins are abundant in many trout streams and their abundance is a good sign that the stream is healthy because they are even more sensitive to pollution and habitat degradation than trout. Nevertheless, trout anglers tend to regard them with suspicion, assuming that a fish so ugly must be harmful to their beautiful trout.

RULING PERCHES

Sea basses (Serranidae) and *temperate basses* (Percichthyidae) are two rather generalized families of the great (seven thousand species) order

black and yellow rockfish

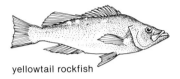

yellowtail rockfish

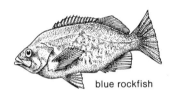

blue rockfish

copper rockfish

kelp rockfish

Figure 5-16. Rockfish are important commercial fishes in both the North Atlantic and North Pacific oceans, often sold in markets as "red snapper." Curiously, there are over sixty species in the Pacific, including the five shown here, but there are only two in the Atlantic.

Perciformes. The order includes many highly specialized families as well, but most are adapted for living in the shallow or surface waters of the oceans or for life in lakes and rivers. The basses, like other perciform families, have:

1. Spines on all (except the caudal) fins.

2. A two-part dorsal fin, the lead half made up of spines.

3. Pectoral fins inserted vertically on the side of the body, with pelvic fins located below them.

4. Ctenoid scales.

5. A swimbladder that serves entirely as an organ of buoyancy (or it is absent).

6. Many distinctive skeletal features.

Most members of the two bass families are moderately deep-bodied predators on other fish. The groupers (Serranidae), for example, are important predators on coral reefs. A number of the serranids start out as females when small but change into males as they get larger, a fact first noted by Aristotle around 300 B.C. The temperate basses are not capable of this sex change and inhabit mainly lakes and estuaries. In North America, important species in sport fisheries are striped bass, white bass, white perch, and yellow bass.

Sunfishes and black basses (Centrarchidae) are only about thirty species and are native only to North America. The family includes such important sport fishes as largemouth and smallmouth bass, bluegill, redear sunfish, and black and white crappies. These species are also very abundant and important predators in freshwater ecosystems of North America. Increasingly, they (but especially largemouth bass) are important in other parts of the world where they have been introduced, often to the detriment of native fishes. The centrarchids are related to the previous two families of basses and are separated by skeletal differences only an ichthyologist could appreciate.

Perches and darters belong to a freshwater family (Percidae) that contains over one hundred thirty species. Ninety percent of them are confined to eastern North America, the rest to northern Eurasia. Walleye, sauger, yellow perch and their European equivalents are highly prized as game and food fishes. One of the most prized is the walleye, so named because of its exceptionally large eyes. These eyes are lined with a special light-gathering layer (tapetum lucidum) that makes it possible for them to hunt other fish under very dim light conditions. When a light is flashed on walleye, the layer gives them "eye-shine," just as many night-active mammals have. Most abundant of the perches are the darters, small, bottom-dwelling inhabitants of streams. Many acquire brilliant body colors during the breeding season and so are finding some favor with coolwater aquarists.

Drums (Sciaenidae) or croakers are heavy-bodied clam (and other invertebrate) crunchers, with strong pharyngeal teeth. Most live in bays, estuaries, and rivers where clams are abundant and where they are much sought after as food fish, especially species like spotted seatrout, spot, weakfish, queenfish, and corvina. The name **drum**, as mentioned earlier, comes from the loud drumming noises they make by vibrating their swimbladders, as part of the courtship ritual. To receive the sounds they have exceptionally large ear stones (otoliths).

Surfperches (Embiotocidae) are a small (twenty-three species) family of fishes abundant in the shallow waters of the north Pacific Rim, from Baja California to Japan. One species, the tule perch, is found only in fresh water in central California. In appearance, they seem to be a more or less standard deep-bodied perciform fish, much like the sunfishes. Their method of reproduction is remarkable, however, because they are livebearers. The young develop to a large size in the female, getting their nutrition from her body fluids, which are absorbed through enlarged dorsal, pelvic, and anal fins. The young are so large and numerous when born that the mother has a difficult time swimming and in the last week or so before birth must spend her time hiding in vegetation. The males in some species are sexually mature when born and begin courtship almost immediately. The females, if receptive, can store sperm for a long period of time before allowing their eggs to be fertilized.

Cichlids (Cichlidae) are the dominant fishes in tropical lakes. About seven hundred species have been described, five hundred from the Great Lakes of Africa alone, but there are probably many more. In these lakes, they form "flocks" of extremely specialized species. For example, some species scrape algae from rocks, others pick small invertebrates from the algae with forceps-like teeth, others sift insect larvae from sand, and a few live on scales jerked from other fishes. The specialization is made possible by the unique and complex structure of the pharyngeal teeth and associated muscles, combined with highly specialized jaw teeth. Essentially they have two sets of specialized jaws, one for obtaining the food, the other for processing it. The success of the cichlids is also attributable to the high degree of parental care they exhibit and the fact that so many species are at least partially herbivorous.

Cichlid interactions with humans are a mixed success for the family. Tilapia cichlids are a favored food fish that have been spread over much of the tropical world by humans because of their hardiness in ponds and their ability to eat just about anything. However, the introduction of predatory Nile perch into Africa's Lake Victoria has apparently resulted in the extinction of two hundred species of native cichlids.

Wrasses (Labridae) and *parrotfishes* (Scaridae) are two closely related families of reef fishes. There are about five hundred species of wrasses, eighty of parrotfishes, although the parrotfishes make up in bulk what they lack in numbers. Wrasses are small, cylindrical fishes with small mouths containing tiny "buck" teeth. They are brightly colored "pickers" of small invertebrates, including parasites from the skins of other fishes. Fishes literally line up at spots where a cleaning wrasse is active, waiting to have parasites and loose tissue removed or to have wounds cleaned. Wrasses and parrotfishes have the most complex mating systems of any vertebrates, with several types of males and individuals changing sex on short notice.

Gobies (Gobiidae) are second only to cyprinids in having the most (eight hundred–plus) species in a family. They occur worldwide in fresh and salt water, especially in tropical areas. As a group, they are easy to recognize because they are rather elongate with blunt heads, bulging eyes, rounded tails, and, most important, pelvic fins united to form a sucker. They have adapted to habitats in which few other fish are found, such as streams on remote islands, cracks and crevices in coral reefs, and the burrows of invertebrates on mud flats. Some can breathe air and will travel across land for short distances. Best known for this are the tropical mudskippers, which will actually climb out of the water in search of food. Most gobies are small and one Phillipine species (*Pendaka pygmaea*) is the world's smallest vertebrate, becoming mature at 6 to 12 millimeters long.

Tunas and mackerels (Scombridae) consist of only forty to fifty species but they are among the most valuable commercial fish species. They are also the top predators of the tropical and subtropical oceans and have extraordinary adaptations for a life of continuous swimming. They feed on squid, herring, and other fishes, which they surprise with a high-speed approach in schools. Their bodies are beautifully streamlined and the skin smooth, with tiny imbedded scales, and an iridescent hue. The dorsal fin

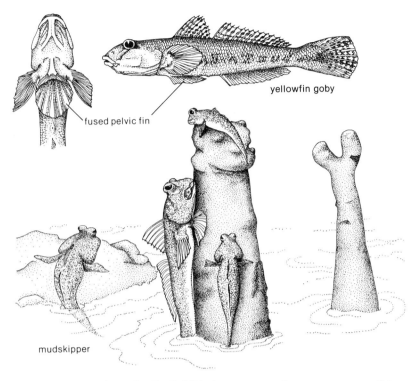

yellowfin goby

fused pelvic fin

mudskipper

Figure 5-17. Gobies (family Gobiidae) are among the most common fishes in the world, but most are small and many occupy unusual habitats. *Top:* Yellowfin goby, showing the typical goby suction disk created by the fusion of the two pelvic fins. *Bottom:* Mudskippers, which regularly emerge from the water in tropical mangrove areas to feed on land.

has stout spines for defense, but they can be lowered into a groove along the back for rapid swimming. The tail is deeply forked or lunate, so it can be moved back and forth at rapid speeds by the powerful body musculature. To increase the efficiency of the muscles, tunas have a special circulatory system that conserves much of the body heat, making them "warmblooded." The swimbladder is reduced or absent, to make diving for prey easier, and water is rammed, rather than pumped, across the gills for respiration. If a tuna stops swimming, it suffocates.

FLATFISHES AND FUGU

Flounders and soles (order Pleuronectiformes) are easily recognizable because they are the only group of fishes that spends their life lying on

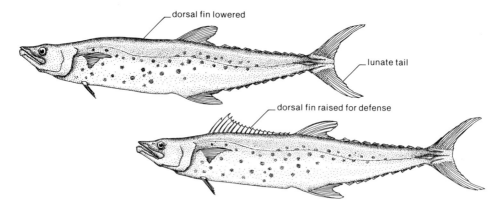

dorsal fin lowered

lunate tail

dorsal fin raised for defense

Figure 5-18.   The sierra mackerel is an example of the fast-swimming members of the family Scombridae, which includes the tunas. When the fish is cruising at high speeds, the spiny dorsal fin fits into a slot on the back to make the fish more streamlined. In the presence of predators, the spines are erected for defense.

one side. The side that lies on the bottom is white and eyeless, whereas the other side is pigmented, with both eyes. Flounders start out as rather ordinary larvae, but as they settle to the bottom an extraordinary transformation begins. The eye on one side of the head migrates to the other side, involving a considerable internal rearrangement of nerves, blood vessels, muscles, and bones. The mouth on many species also acquires a twist, to improve the efficiency of bottom feeding. Curiously, some species of flounders have their eyes on the right side of the body, some have them on the left, and some have them on either side. There are over five hundred species of flatfishes, many of them valuable commercially.

Puffers, boxfishes, and porcupine fishes (order Tetraodontiformes) are one hundred fifty species of slow-swimming, heavily protected fishes with strong beaks. The beaks are used for shearing off coral and other hard-shelled invertebrates. To make themselves less palatable to predators, most members of this group can puff themselves up with large quantities of water, which not only makes them physically larger but causes the spines in their skin to bristle. As if this were not enough, some species are also poisonous. This has not kept them from being prized food fishes in Japan, as fugu. Fugu is not poisonous if properly prepared so only licensed cooks are allowed to do so. Because cooks make mistakes on occasion, the flavor of the fish is enhanced by the danger involved in eating it.

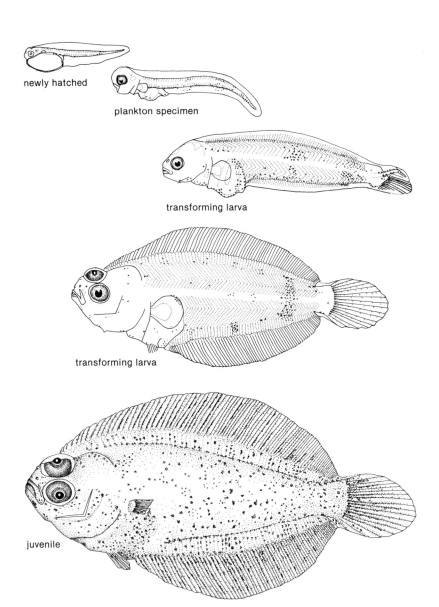

newly hatched

plankton specimen

transforming larva

transforming larva

juvenile

Figure 5-19. Flatfish, such as the butter sole, start life as normal fish larvae, swimming upright, with eyes on both sides of the body. Gradually, one eye migrates to join the other on one side of the body and the fish begins a life of bottom living, lying on its blind side.

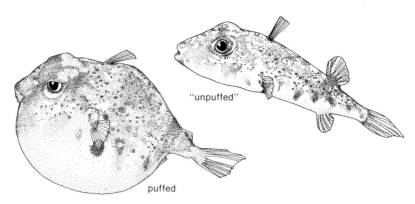

"unpuffed"

puffed

Figure 5-20. Puffers are slow swimmers that make themselves difficult for predators to eat by puffing themselves up with water.

PROJECTS

The best way to learn how to appreciate the diversity of your own local fish fauna is to acquire a local guide or key to the fishes, which are increasingly available (see chapter 16). Catch fish by whatever legal means you can and key them out, first to family and then to species.

CHAPTER VI

Ecology

I have spent a great deal of time under water, watching fish. The observations I make are carefully recorded on special forms and converted into data. The data are analyzed, summarized, and eventually turned into one of the arcane documents known as a scientific paper. The publications enable me to call myself a fish ecologist but they do not convey how enjoyable the underwater excursions are. They also do not allow me to record the myriad of other experiences I have had, such as watching how light and water together create lovely, shifting patterns on a lake or stream bottom, how a larval mayfly scurries across a stone in fast water, or how a lamprey moves a large rock to excavate a nest. I call myself an ecologist, but my motivations and even my methods are nearly the same as those whose work is the foundation of ecology, the natural historians.

Ecology is the study of the relationships between organisms and their environment (including other organisms). It is now a respectable branch of the biological sciences and is becoming increasingly sophisticated and quantitative in its methods. Its roots are in natural history, which is usually interpreted as the observational study of wild organisms. Most of the great biologists of the nineteenth century, such as Louis Agassiz and Charles Darwin, were natural historians and their works are largely descriptive rather than quantitative. Early in this century, natural history fell into disrepute, as the "hard" sciences of physics and chemistry developed, together with more quantitative approaches to biology, especially in genetics. It did not help that the prose of many natural his-

Figure 6-1. Adult European kingfisher feeding a minnow to its young, in a nest hollowed in a riverbank.

torians tended to be flowery and that arguments from authority, rather than evidence, were common. Nevertheless, the study of nature has always had its appeal and the development of the science of ecology, with its own esoteric jargon, was one way to study the biology of wild organisms and communities while maintaining scientific respectability.

Today both ecology and natural history are flowering, as shown by the large natural history sections of most bookstores and the enormous selection of publications available in scientific libraries. The gap between natural history and ecology (if indeed it ever existed) has also been bridged repeatedly by the wonderfully descriptive prose of scientists such as Rachel Carson and Loren Eiseley and by the lucid accounts of scientific studies found in magazines such as *Natural History* or *Audubon*. Ecology is still one of the most accessible of sciences to the nonscientist and an area in which amateur naturalists can still make contributions.

The purpose of this chapter is to provide some general background in fish ecology, by describing the responses of fish to dominant environmental factors. Although fish, like all organisms, have to respond to many factors simultaneously, for convenience the factors are divided here into physical, chemical, and biological factors.

PHYSICAL FACTORS

The most obvious and most studied physical factors affecting fish are temperature, light, water currents, pressure, and substrate (bottom type and cover). Fish also respond to other more subtle factors, such as the earth's magnetic fields.

TEMPERATURE

With the exception of a few large "warm-blooded" tunas and sharks, body temperatures of fish are always close to those of the environment. The main reason for this is that gills of fish constantly put the blood of fish in close contact with the water, for efficient exchange of oxygen; this also results in continuous loss to the environment of heat generated by muscular activity. Despite this problem (from a mammalian perspective), different species of fish are adapted for life at a wide range of temperatures. Some antarctic fishes spend their lives in water just slightly above freezing and may even rest on submerged ice. Some desert pupfishes inhabit water that reaches 42° C. Furthermore, many species of fish are capable of living under a wide range of temperature conditions and of adjusting to wide daily fluctuations in temperature. Despite this adaptability, each species has a definite range of temperatures within which it can live, and each individual, at any given time, has an even more limited range than the species as a whole.

The temperature range for each individual is determined largely by the temperature of the water in which it has been living (**acclimation temperature**). At any given temperature, a fish, depending on its acclimation history, may live or die, be active or torpid, migrate or stay put, or spawn or not spawn. The particular range of temperatures a fish can withstand may also depend on its life history stage. Generally, immature stages are less tolerant of temperature extremes and fluctuations than are adults. Obviously, the effects of temperature on fishes are many and complex, so here we will examine only its effects on distribution, movements, and growth and survival.

Distribution.   The importance of temperature in determining the distribution of entire faunas is indicated by the ease with which coldwater and warmwater fish faunas can be recognized. Species and even families and orders of fishes differ considerably among the polar, temperate, and tropical regions. Salmon, trout, smelt, and their relatives, for example, are confined to the cold waters of the world. In the North Pacific Ocean, the dominant pelagic (open water) predators are salmon, which are most abundant in waters of 3°–9° C. Farther south, the tunas and mackerels become the dominant pelagic predators, and are most abundant in waters of 15°–21° C. The waters off California are transitional between these two temperature regions, so major changes in pelagic fishes often accompany minor changes in temperature. When water temperatures off southern California change from an average of about 15° C to 16°–17° C, many species of subtropical fishes, such as barracuda, yellowtail, and bonito suddenly appear in the catch of fishermen that in other years may be catching salmon.

Decreases in temperature can also have dramatic effects. In 1882, there was a massive die-off of tilefish off the coast of New England that followed a drop of a few degrees centigrade of the bottom water in which they lived. It took the tilefish population more than thirty years to recover from this natural disaster.

LIFE STAGE	WATER TEMPERATURE REQUIREMENT FOR GROWTH (°C)	LOCATION OF PACIFIC SALMON LIFE STAGE		
		freshwater	estuary	open sea
egg	3.5 - 13			
alevin (sac fry)	5 - 13			
fry	5 - 17.5			
parr	5 - 18			
smolt	5 - 19			
adult	5 - 19			

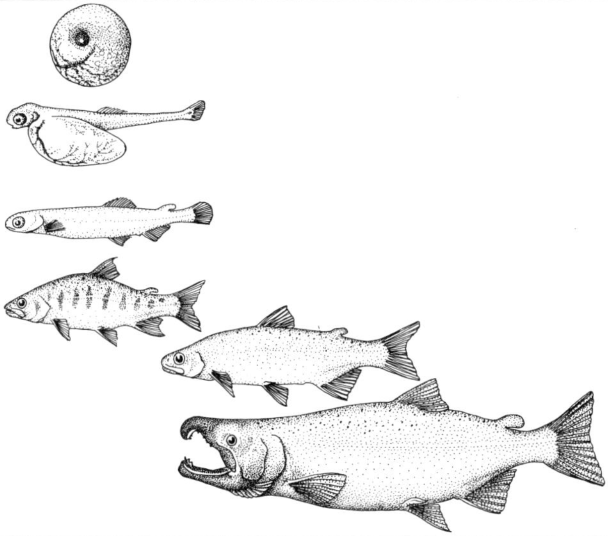

Figure 6-2.   Like most fishes, the life stages of Pacific salmon have different temperature and habitat requirements.

Movements.   Many species of fish are capable of detecting temperature changes as small as .03° C, so it is not surprising to find that fishes often key their daily and seasonal movements to temperature. Because each species has an optimal temperature range at which it functions most efficiently, movements are often aimed at staying within this temperature range. Thus tuna move north as the oceanic waters warm up in the spring and south as they cool in the autumn. In the Atlantic Ocean, American shad prefer water of 13°–18° C, and their seasonal movements up and down the coast are keyed to these temperatures. In temperate lakes, fishes tend to move into shallow water in spring as the water warms up (often for spawning, as the young grow faster at warmer temperatures). In the fall, these fishes move into deep water, where the water will be a uniform 4° C. Water is most dense at this temperature and so stays below the less dense layer of colder surface water. Temperature change is often an important cue for spawning migrations; for suckers a rapid increase of water temperature in their spawning streams in the spring is typically the cue that starts migration to spawning areas.

Growth and survival.   Growth rates of fishes are highly dependent on temperature. In the laboratory, it can be demonstrated that each species has an ultimate preferred temperature and that this temperature is the temperature at which it grows the fastest (if food is in unlimited supply). For rainbow trout this temperature is 18°–21° C, for brook trout, 14°–18° C, for walleye, 20°–23° C, for common carp, 27°–32° C, and for bluegill sunfish, 30°–32° C. In the wild, fish are usually found at temperatures above or below the optimum for many reasons, such as greater abundance of food or more cover from predators in the less optimal areas. If given a choice, however, they will live at temperatures as close to the optimum as possible because this allows them to grow faster. This is advantageous because the bigger a fish is, the fewer predators it will have. More rapid growth may also allow fish to reproduce at younger ages or to produce more eggs.

LIGHT

Fish live under a wide variety of light conditions, from brightly lighted surface waters to the absolute darkness of caves and ocean abysses. Because light varies on a regular daily and seasonal basis, it is perhaps the single most important environmental cue that fish use to regulate their daily and seasonal activity patterns. Like temperature, light is a complex factor, so fish respond not only to its periodicity but to such aspects as its direction (polarity) and intensity.

Periodicity.   Most fishes have distinct daily and seasonal activity patterns that are cued by light. Many predatory fishes, for example, have peaks of feeding at dusk and dawn, when light levels are high enough to see prey but low enough to make a careful predator hard to see in return. In streams, trout show strong peaks of feeding at dusk and dawn, largely in response to the availability of their invertebrate prey, which are most active at night. The prey start to become active as light levels drop. In complex fish communities, such as those found on coral reefs and many warmwater lakes and streams, there are usually both day-active and night-active fishes, which thereby avoid direct competition with each other. This is epitomized by the situation in fresh water where catfishes are often the most active fishes at night, whereas minnows and characins are most active during day. These fishes, although very different in appearance, have common ancestry and many internal similarities.

In oceans and large lakes, many pelagic fishes show strong daily vertical migrations of as much as 500 meters, moving towards the surface at dusk and down again at dawn. Such movements are cued by light and the main reason for them is to follow vertical migrations of zooplankton, upon which the fishes feed. Avoidance of larger predatory fishes may also be a factor in these migrations. Horizontal movements of fish with changes in light are also common. In lakes, streams, and coastal marine waters, many species move into shallow water at night, either to eat or to avoid being eaten. Thus in lakes, black crappie are often found offshore during the day, where they feed on zooplankton. As light levels decrease they may move inshore and switch to larger prey, mostly insect larvae and small fish.

The reproductive cycles of fish are usually cued to seasonal changes in day length. For spring spawning fish, such as rainbow trout, the gonads mature and spawning migrations begin as day length increases. For fall spawning fish, such as brook trout, the same effects are caused by decreasing day length. Knowledge of this phenomenon has permitted the operators of trout hatcheries to control the time of spawning of their' trout by controlling light cycles in hatcheries.

Direction. The polarity of light is an important cue for fishes that make long migrations. They can use the sun for navigation, much as human navigators do, using the position of the sun as the reference point. Thus Arthur Hasler and his students found that, in Wisconsin lakes, white bass could find their way back to their spawning grounds (from which they had been removed) if the sun was shining, but could not if the sky was overcast.

Intensity. For fishes that make vertical migrations, the intensity of light is often the most important cue that tells them where they are in the water column. In the ocean some species will stay in a rather limited range of light intensity and stay close to this range when moving up and down the water column. Light intensity also influences predator-prey interactions because most fishes rely heavily on vision for finding prey and for detecting predators. Prey size tends to increase as light levels decrease. In addition, schools of fish normally break up at dusk, because

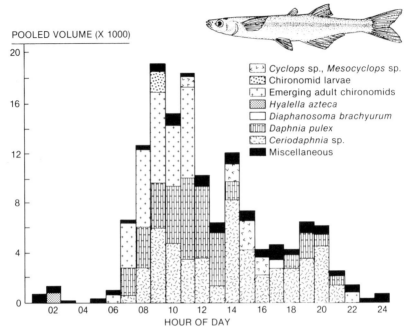

Figure 6-3.   The inland silverside in Clear Lake in California is typical of many fishes that rely on vision for finding their food. It feeds mainly during the day, with a peak in the morning after an inactive night. Note also that different types of prey are consumed at different times of day. For example, adult chironomid midges are taken only in the morning, when they are emerging from the water.

their structure is maintained by visual contact among the members. The schools form again just before dawn.

When suspended or dissolved material permanently reduces penetration of light into water, the dominant fishes are those that do not rely heavily on vision for feeding: bottom grubbers such as carp or tactile predators such as catfish. In Africa and South America, numerous species of fishes evolved that use electrical fields to find their way about the perpetually turbid conditions of tropical rivers.

WATER CURRENTS

No matter what type of environment a fish lives in, it is going to encounter water in motion: currents in streams, oceanic currents, tidal currents, wave action, and even the slight turbulence created by its own swimming movements and those of other fish. The prominence in fishes of the lateral line system, a system designed mainly to detect turbulence, attests to the importance to them of information on currents. In streams, water velocity is one of the best predictors of species distributions. Swift waters contain mainly streamlined, fine-scaled fishes such as trout, whereas sluggish waters are more likely to contain deep-bodied, large-scaled forms such as sunfish and carp. Likewise, turbulent waters favor small, streamlined fishes that can live under rocks or in crevices to avoid the strong currents. Thus the shallow wave zones of lakes are often in-habited by the same fishes (such as sculpins and dace) that occupy the riffles of nearby streams.

Currents are a major feature of oceans, much like the rivers of con-tinents. Fishes take advantage of them, especially for spawning. The adults of many species, such as herring, migrate against the surface cur-rents to a spawning area, so the young will be carried by the currents back to the adult feeding grounds. This is also what causes salmon to swim upstream to spawn.

PRESSURE

The swimbladder of bony fishes is a marvelous adaptation because it allows a fish to stay suspended in the water with no effort. It does have a major disadvantage, however, in that it greatly limits the ability of fish to move up and down the water column because changes in pressure cause changes in swimbladder volume. For each 10 meters a fish swims upward, the gas in the swimbladder doubles in volume! Most bony fishes can only adjust the volume of their swimbladder slowly and can tolerate only about a 25 percent increase in swimbladder volume in a short period of time. This means that most fish tend to stay at a rela-tively narrow range of depths. Anglers who catch fish in deep water know that if they pull a fish to the surface and release it, it will float helplessly unless the swimbladder is deflated. Fishes that migrate up and down through the water column apparently have their swimbladder vol-ume adjusted for the top of their vertical range; this means that they must swim actively to keep from sinking at the bottom of their range.

SUBSTRATE

Most fishes have swimbladders or other means of buoyancy so are the-oretically "free" of the bottom. In fact, most fishes do associate with the bottom or some substrate in one way or another. Even pelagic fishes often orient to floating objects. Substrates, of course, are sources of food and shelter and provide the diversity of habitat that permits many species to live together. As the next chapters will show, many species can survive only on or in very specific substrates, such as beds of aquatic plants or underneath large boulders. Other species of fish choose very specific substrates for spawning: salmon and trout seek out gravel bottoms in riffles; yellow perch, beds of aquatic plants; and some cichlids, muddy bottoms for nests.

CHEMICAL FACTORS

Water is the "universal solvent" and fish are extremely sensitive to the diverse chemicals dissolved in it, including the many pollutants that have been added by humans. Here only three of the more important chemical factors will be discussed: dissolved oxygen, salinity, and pH (acidity).

Oxygen.   Air is a uniform mixture of oxygen and other gases, so terrestrial animals rarely have to be concerned about the amount of oxygen available to them. In water, not only is the amount of oxygen available much less than that in air, but it varies from place to place and from time to time. Effects of dissolved oxygen levels on fish distribution are often hard to separate from effects of temperature, because cold water can hold more oxygen than warm water. Thus water at 1° C is saturated with oxygen at about 14 parts per million (ppm), whereas water at 30° C is saturated at less than 8 ppm. The situation is further complicated by the fact that at lower temperatures, fishes have lower metabolic rates and hence less demand for oxygen than they do at higher temperatures. Thus yellow perch at 4° C will not die until oxygen levels drop well below 1 ppm, but at 20° C they will die when oxygen levels fall below 2.25 ppm.

   Lethal oxygen levels are only part of the problem for fishes, however, because well before those levels are reached, critical oxygen levels are reached, at which fish must restrict their activity in order to lower their oxygen consumption rate. If critical oxygen levels are maintained for a long period of time, growth and reproduction in fish can be severely reduced. This is one of the more subtle effects of sewage pollution, because decaying organic matter uses up dissolved oxygen. One of the reasons common carp are often so successful in such polluted conditions is that they are unusually tolerant of low oxygen levels and can be active at levels that would be lethal to many other species. In tropical waters, where oxygen levels are often naturally low due to organic matter in the water, many fishes have evolved the capacity to breathe air.

Salinity. The importance of salinity as an ecological factor is reflected in the distinctiveness of freshwater and saltwater fish faunas. The dominant group of freshwater fishes, for example, the minnows, characins, and catfishes, has only a few species that even enter salt water at low salinities, whereas most marine families contain at best only a handful of species that will enter fresh water. Most fishes in both environments occur in a relatively small range of salinities (are **stenohaline**), although some fishes are adapted for living in a wide range of salinities (are **euryhaline**). Stenohaline freshwater fishes avoid salinities higher than 3 to 5 parts per thousand (ppt), whereas stenohaline saltwater fishes avoid water that is less than 30 to 34 ppt (sea water is 35 ppt). Euryhaline species occur in areas where salinities naturally fluctuate, in estuaries, in lakes in arid areas, and in the intertidal zone of the ocean. In marine fishes, the ability to withstand low salinities may be greater in early life history stages because this allows them to use estuaries as nursery areas. Estuaries are bays where salt water and fresh water mix and where nutrient levels are high as a consequence. The young fish in the estuaries can grow rapidly despite the stress of fluctuating salinities; thus species such as menhaden and many croakers are found in estuaries mainly as young. Some euryhaline predatory fishes, such as striped bass, will spend much of their adult life in estuaries to take advantage of the abundance of such small fishes as a food source.

Most fishes of the intertidal zone are moderately euryhaline, because tide pools and small bays may become diluted with rain water or have increased salinities due to evaporation. Only a small number of species in this zone can tolerate the extreme conditions that occur in the most exposed portions, however. In California, the longjaw mudsucker (a goby) can survive salinities ranging from 12 to 83 ppt, whereas on the east coast, the sheepshead topminnow (a killifish) can live in water ranging from fresh to 142 ppt. Some of the desert pupfishes, relatives of the sheepshead topminnow, also tolerate a similar wide range of salinities.

pH. pH is a measure of the hydrogen ion concentration in water and reflects water's acidity. Water that is neutral has a pH of 7, whereas acid water has values less than 7 and alkaline water has values higher than 7. The pH of the ocean is buffered by all the salts so it is always con-

stant, but fresh water shows wide variation in pH. Most freshwater
fishes can live in waters with pH values ranging from 5 to 9, although
sudden transfers between waters of widely differing pH values can be
lethal. Some fishes can survive at slightly greater or lower pH values,
but in general water with pH values greater than 10 or less than 4 will
be fatal to fish.

Within the 5 to 9 pH range in which most fish live, the effects of pH
are subtle. In general, the higher the pH, the greater the productivity of
the water and the faster the growth rates of the fish. The most productive lakes are those that are slightly alkaline (pH around 8). This is also
true of streams. For example, in a classic study, Winnifred Frost found
that the growth rates of brown trout in an Irish stream were much
higher in the lower reaches of the river (pH 7.8–8.0) than in the more
acid upper reaches (pH 5.6), even though flows and other conditions
were about the same.

Despite high productivity, alkaline waters greater than 8.5 pH will
support only a few fishes adapted for living in such waters. One such
species that has been widely planted in alkaline lakes of the western
United States is the Sacramento perch, a relative of the sunfishes. Were
it not for such plantings, it would undoubtedly be on the endangered
species list, because it is now quite rare in its native Sacramento Valley
of California. Another fish planted for sport fishing in alkaline lakes is
the Eagle Lake rainbow trout, which evolved in a large alkaline lake in
California, where pH levels may reach 9.6.

Although there are a number of fishes that will thrive in highly alkaline waters, this is not the case for highly acidic waters, in part because high acidity is often associated with high concentrations of heavy
metal ions, which are toxic in their own right. The inability of fishes
(and other organisms) to tolerate high acidity is the reason acid rain is
of such concern. As the accumulation of acid rain causes the pH of lakes
and streams to drop, the fishes disappear. The first sign of trouble is
decreased growth rates and as the problem worsens, the fish become
unable to reproduce. Lower levels (less than 4.5) may kill the fish directly. There are numerous mountain lakes in the eastern United States,
Canada, and northern Europe with no or reduced fish populations because of acid rain. The acidity of these lakes makes them as clear as a
chlorinated swimming pool and just as supportive of life. A similar situation exists with streams polluted with acidic water that drains from
coal and metal mines.

BIOLOGICAL FACTORS

Physical and chemical factors have an obvious effect on the distribution and ecology of fishes. Biological factors are often equally important but are much harder to understand because of their subtlety and complexity. For convenience, biological factors can be divided into predator-prey interactions, competition, and symbiosis. Although only the interactions between species will be discussed here, it should be kept in mind that similar interactions within a species (e.g., cannibalism) are also important, especially in the regulation of population size.

Predator-prey interactions.   Most fishes are both predators and prey. The vast majority of fish live with the constant threat of being eaten by predatory fishes larger than themselves. Indeed, much of the morphology of fishes can be explained as evolutionary attempts to become the most efficient predator possible while reducing the probability of being eaten by someone else. A stickleback, for example, has locking spines on its fins and bony plates on its sides to protect itself, and also has a small sucking mouth and maneuverable fins that allow it to feed on the small invertebrates it favors. Requiem sharks, such as white sharks, are renowned for their adaptations for feeding on fishes and marine mammals, yet the fact that they give birth to large-sized young is best interpreted as a way to reduce predation on these young.

Because both predators and prey exist in nature, balance must be achieved through the juggling of evolutionary and environmental forces. The delicacy of this balance can be observed when humans upset a system through such means as introducing new species. Thus the introduction of Nile perch into Lake Victoria resulted in eradication of numerous species of small native fishes. Similarly, the accidental introduction of sea lamprey into the Great Lakes caused a crash in the populations of the large fish they fed upon. Because predators consume large amounts of fishes also desired by humans, it is often assumed that reduction of predator populations in natural systems will increase the number of fish available to fisheries. Such control programs have rarely worked. For example, on the Pacific coast, large numbers of predatory fishes were removed from lakes used by juvenile sockeye salmon as nursery areas. This greatly increased the number of young salmon going out to sea, but the average size of these fish was much smaller than before, because the juveniles were competing with each other for the limited food available in the lake. Once at sea, these small juveniles presumably were more vulnerable to marine predators, because no increase in the number of adult salmon returning a few years later could be detected.

At times, the abundance of prey may regulate the abundance of a predator. Off California, brown pelicans produce the most young in years when northern anchovy, their principal prey, are most abundant. In ponds where bluegill sunfish are abundant, bluegill may keep the number of largemouth bass low by preying on eggs in bass nests. This is especially likely to happen if anglers remove the largest bass, which might keep bluegill populations in check through predation.

Competition.   Competition occurs when two organisms (or species) require some resource for their survival and that resource is in short supply. Usually, the limited resource is food or space. Introduced brown trout will aggressively displace native brook trout from the limited number of places where there is both cover and easy access to food drifting down a stream; the brook trout are then forced to use more exposed areas where food is less available and they are more vulnerable to anglers and other predators. Eventually, brown trout may completely eliminate brook trout from a stream through this competitive mechanism.

Such **direct competition** is actually rather uncommon; more common is **exploitative competition,** where one species uses a resource more efficiently than another and eventually may make it unavailable to the other species. For example, R. Johannes and P. A. Larkin showed that redside shiners (minnows) were much more efficient at feeding on scuds (small shrimp-like invertebrates) than were small trout. The shiners could seek out scuds in beds of aquatic plants, whereas the trout had to wait for them to come out. When shiners were introduced into a trout lake, they quickly reduced the abundance of scuds, which had been the principal food of small trout. The growth and survival of small trout became greatly reduced and the trout were forced to switch to feeding mainly on insects that fell on the water's surface. It is important to note in this example that the trout did manage to survive by switching their food source. Indeed, it is likely that if the trout–shiner lake was isolated long enough, the trout would evolve means to use the terrestrial insects (or other food sources) more efficiently. This evolutionary mechanism may be largely responsible for the diversity of body shapes and the many feeding specializations we see in communities of fishes. They allow fishes to avoid competition with other species.

SYMBIOSIS

Symbiosis means "living together" and is used to cover situations where two or more species live in close contact with one another and interact in ways other than predation or competition. Symbiosis is divided into three categories, which blend into one another: mutualism, commensalism, and parasitism.

Mutualism. Mutualism is an interaction between individuals of different species in which both species benefit. In fishes, its most common occurrence is in species that school together. Joint schooling allows bigger schools, which apparently reduces predation (see chapter 5). For example, in streams and lakes of the eastern United States, several species of minnows will school together, often with different species occupying different parts of the school. Mutualism can also involve very different kinds of organisms. Anemone fish that use an anemone for cover and protection of eggs may feed the anemone small fish on occasion. Some species of goby live in the burrows of shrimp, a behavior the shrimp encourage because the gobies will warn the shrimp of the approach of predators.

One of the most interesting forms of mutualism is **cleaning behavior,** where one species picks off and feeds on the parasites and diseased tissue of another species. This behavior has been observed in a wide range of species but the cleaning wrasses and gobies found on coral reefs are most studied. The reason for this is that such fishes often set up cleaning stations (which they advertise with their bright colors and special "dances"), where larger fish literally line up to be cleaned and go into a special trance-like state when being cleaned (for further discussion see chapter 14). Sometimes the desire of a fish to engage in cleaning behavior creates some interesting conflicts with other behaviors. For example, I once observed a small male bluegill vigorously patrolling and defending its nest, located on the edge of a nesting colony of bluegill. Among the fish it would periodically chase away were juvenile smallmouth bass about half its size. I was greatly surprised, therefore, when one of the bass adopted the rigid posture of a fish wanting to be cleaned and entered the nest—and the bluegill proceeded to clean it, slowly picking over one side of the bass. When it was through, it chased the bass out of the nest. This conflicting behavior was repeated several times in a short period.

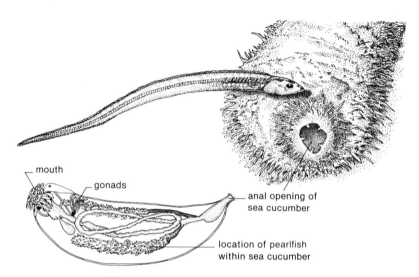

Figure 6-4.   Pearlfish are parasites on sea cucumbers. They enter the cucumber through the anus and penetrate into the body cavity, where they feed on the gonads.

Commensalism.   This form of symbiosis occurs when two species live together but only one species gains from the relationship, while the other is not harmed by it. Many species of gobies, for example, live in shrimp burrows but, unlike the mutualistic gobies previously mentioned, do not deliberately warn the shrimp of approaching predators. On coral reefs, cornet fishes may use large parrotfish as mobile cover in their search for small fishes to eat. The remoras or sharksuckers are a whole family of fishes (Echeneidae) adapted for living commensally with sharks, whales, and larger turtles. A remora has a large sucker on top of its head that permits it to attach to the bottom of a passing host, which then carries it to new sources of food. Remoras are actually a somewhat ambiguous example of this phenomenon because their presence certainly costs the host some energy (typical of parasites). However, some remoras have been observed to act as cleaners on their hosts as well, removing other external parasites.

Parasitism.   All fishes are host to a variety of invertebrate parasites but fishes that actually act as parasites are few. The pearlfishes (Carapidae) are one of the few; they are small, elongate marine fishes that parasitize sea cucumbers, large sluggish invertebrates. A pearlfish enters a sea cucumber's gut through its anus and once inside may penetrate into the body cavity, to nibble on the gonads. The sea cucumber will not be killed by the pearlfish but it is likely to have decreased ability to reproduce. Examples of parasitism by freshwater fishes are found in the South American catfish family Tricomycteridae. Some of these small catfishes enter the gill cavities of larger catfishes and feed on the gill filaments and blood. Other species ambush passing fishes to snatch scales from them. Scale snatching is actually accomplished by a wide variety of fishes, but whether or not it is an example of parasitism or of specialized predation is debatable.

PROJECTS

1. If you fish on a regular basis, keep a journal of your catch and record temperature, weather, and water conditions each time you go out. If you live near a lake or stream, make periodic observations of temperature, flow, and other variables, and see how they relate to observations you can make on the fishes, such as spawning runs.

2. pH paper is readily available in drug stores. Using it, test the acidity of local waterways and compare them to the acidity of local rain. See if there is any relationship between fish abundance and pH.

CHAPTER VII

Trout Streams

Coldwater streams are the epitome of flowing waters: clear, pure, and cold, falling and rippling down rocky mountainsides or smoothly meandering through flower-filled meadows. They are defined, in most people's minds, by their ability to support trout and their close relatives, salmon, grayling, and whitefish. Yet a surprising number of such streams were without fish until trout were planted in them by eager anglers, especially in western North America. Even trout streams that already contained native trout now find themselves with one or more additional trout species. Brook trout from eastern North America have been planted in rainbow and cutthroat trout streams of the West, whereas rainbow trout have been planted in brook trout streams of the East. Brown trout from Europe have been planted in streams through-out North America whereas brook trout and rainbow trout have been planted in Europe.

Trout streams also typically contain other fishes besides trout. In coastal waters we find salmon and trout living together, sometimes four or five species in one stream system. Most trout and salmon naturally live with several species of "nongame" fish as well, such as suckers, dace, and sculpins. Such fishes are much maligned by anglers, often be-cause they are abundant when trout are scarce; it rarely seems to occur to the complainers that excessive removal of trout by fishing may have something to do with an apparently unbalanced stream. It is even more likely that the scarcity of trout is due to environmental changes caused by the damming and diversion of the streams, the logging or overgrazing

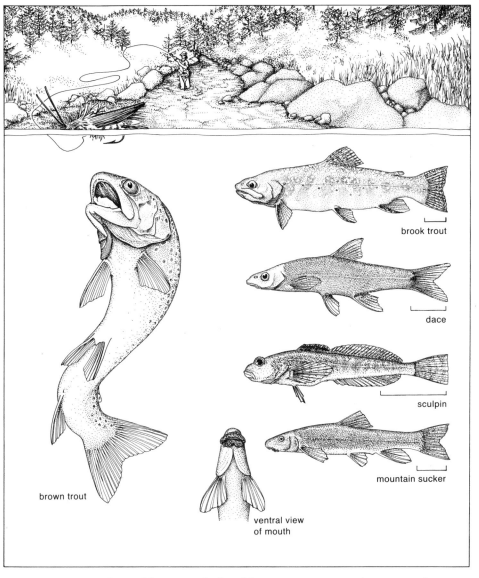

Figure 7-1. Some fish commonly found in trout streams (not drawn to scale; scale line under each fish represents 2 centimeters).

of the stream watershed, or pollution from nearby towns or distant factories.

The purpose of this somewhat convoluted introduction is to make three points:

1. Trout streams are surprisingly complex and delicate environments.

2. Trout and their relatives are very sensitive indicators of environmental change.

3. Even trout streams, the epitome of flowing waters, have been extensively altered by human activity.

All these things, of course, do not keep trout streams from being one of the more accessible and interesting aquatic environments. These points should become even more apparent as we examine some general concepts in stream biology and then discuss the factors that limit the abundance of stream fishes.

SOME CONCEPTS OF STREAM BIOLOGY

Streams change greatly in character from their headwaters to their lower reaches. In mountainous areas, the headwaters are typically turbulent, clear, and cold. As headwater trickles coalesce to form larger creeks and as creeks unite to form even larger streams, the water gradually loses much of its clarity, becomes warmer, and pools and slow runs (glides) become more and more the dominant habitat type. Because headwaters are small and usually flow through heavily forested areas, they are heavily shaded. This helps to keep them cool but it also inhibits algae and other aquatic plants—plants that form the base of most aquatic food chains—from growing in the stream. Trout streams contain a rich assortment of life, however, because there is a constant input of energy in the form of leaves, pine needles, twigs, and other organic debris from the surrounding forest. Much of this material, of course, is washed downstream where it collects in quiet water, but much remains as well, trapped in pools, behind dams of debris, in back-eddies behind rocks, or in the nets some aquatic insects spin to catch small particles.

It is the aquatic insects that make it possible for organic debris to be converted into fish flesh. Some forms, **collectors**, trap bits of organic matter which they consume. Others, **shredders**, cut up and eat leaves

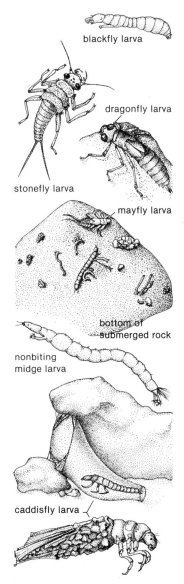

blackfly larva

dragonfly larva

stonefly larva

mayfly larva

bottom of
submerged rock

nonbiting
midge larva

caddisfly larva

Figure 7-2. Common insect
inhabitants of trout streams.
These insects are also found in
trout stomachs.

and other large pieces of debris, with the leftovers becoming more vulnerable to decay by bacteria or drifting downstream to be trapped by collectors. What algae do manage to grow in the scattered patches of sunlight are consumed by **scrapers,** insects that scrape material from the surfaces of rocks and logs. All these insects are eaten by **predators,** which include not only fish, but a variety of other insects.

The number of different types of insects varies with environmental conditions. Shredders are most abundant in streams heavily shaded by deciduous trees. Collectors often are found in extremely high densities below the outlets of lakes because they collect the lake plankton in their nets. Scrapers are most abundant in sunlit sections, so are often dominant in streams too large to be covered with a canopy of trees. In such streams, grazing scrapers often keep the algae growing on rocks cropped down to a very low level; if they disappear, blooms of green algal filaments may quickly cover the rocks. In the McCloud River in California, for example, the principal grazers seem to be large caddisfly larvae, which carry around protective cases of stones and twigs. These larvae begin the process of metamorphosis into the flying adult stage all at once and cease feeding. Within a few days, the rocks become green and slippery with algae, and much more hazardous to the hapless trout angler as a consequence.

Regardless of the way they feed, aquatic insects are the most important food of most stream fishes, either as larvae in the streams or as adults flying above the stream and landing on the water. Naturally, the insects try very hard not to be eaten and have evolved many ways of making it difficult for fish to capture them. Many caddisflies, for example, encase themselves in pieces of debris, both to disguise themselves and to cover themselves with inedible armor. Blackfly larvae glue themselves to rocks in water too fast and shallow for most fishes, whereas net-spinning caddisflies live beneath rocks, covered in part by their nets. Some larvae of mayflies and all stoneflies are extremely flat so they are very hard to distinguish from the rock surface. In addition many aquatic insects are most active at night when fish have a hard time seeing them.

This night-time activity of invertebrates results in a phenomenon known as **drift,** whereby individuals by accident or design release their hold on the bottom and drift downstream, to settle in a new spot. Drift occurs at low levels at all times but increases greatly at night, starting at dusk and decreasing at dawn. Trout and other fishes take advantage of this and feed heavily at dawn and dusk, as most trout anglers know.

The feeding peaks are exaggerated by the fact that terrestrial insects also become more active during the crepuscular hours and hence are more likely to fall into the stream. In addition, many aquatic insect adults hatch from their submerged pupal cases at dawn or dusk. One of the advantages that some nongame fishes have over trout is that they do not have to depend on these daily food peaks. Sculpins, for example, are small flattened fishes that live down among the rocks in fast water and ambush passing larvae. Suckers are browsers with strong flexible lips that suck in algae and fine bottom debris, together with small insects that live in such material.

If aquatic insect communities show striking changes in a downstream direction, so do fish communities. The changes are gradual in most streams. The headwaters may contain only a few small brook trout, which are joined by rainbow or brown trout and a few sculpins as the stream increases in size and temperature. A bit farther downstream small schools of dace may be found on the edge of the riffles, while suckers feed on the bottom of pools. A large piscivorous brown trout or two may be found lurking underneath an undercut bank, while other trout feed on insects drifting into the head of the pools. As the stream warms up and becomes more turbid, trout become less abundant and streamlined minnows and suckers of various species swim about in the quiet water in large schools. By the time the stream has joined with others to become a lowland river, the number and variety of fishes have increased greatly; bass pursue the schools of minnows and deep-bodied sunfish hang suspended in the water next to beds of aquatic plants and fallen logs.

In streams where gradients are high, the transitions between different fish communities can be quite rapid, and distinct groups of fishes will occupy distinct regions of the stream. Such regions are often called **fish zones**. In the San Joaquin River of California, for example, the water originates on the steep west side of the Sierra Nevada. The high mountain waters are occupied exclusively by trout, forming a Trout Zone. In the foothills, the trout are largely replaced by nongame fishes because of high water temperatures, forming the Squawfish-sucker-hardhead Zone. As the water spills out onto the valley floor, the foothill fishes are joined by a variety of fishes adapted for life in warm sluggish water, forming the Deep-bodied Fishes Zone. These warm water environments generally have many more fish species than trout streams and will be discussed in the next chapter.

LIVING IN AN UNCERTAIN ENVIRONMENT

Trout streams are encountered by humans mainly on nice sunny days, after the opening of trout season in the spring and before its close in the fall. Such encounters give the impression that streams are fairly constant entities, so anglers are often surprised when a stream that provided excellent fishing one year does not a year or so later, even though it has not been heavily fished. The reason for this unpredictability is often the timing and duration of winter and spring floods, which can wash fish downstream, flush developing embryos from the gravel in which they have been buried, and crush sculpins and other small bottom fish with rolling boulders. In one small trout stream, which I have spent a number of years studying, I estimated in one year that over 1,200 fish of five species lived in one 35-meter-long section; two years later the estimate was only 55 fish. The first estimate was made in a summer that followed two consecutive winters virtually without flooding, whereas the second estimate was made following a winter of exceptionally heavy flooding. Many other factors also limit the abundance of fish in streams, such as temperature, water chemistry, and cover.

Temperature is particularly important because it regulates the metabolic rates of fishes, so each species has a range of temperatures in which it is likely to thrive. Trout and sculpins seem to have especially narrow ranges but both do best in moderately cool water (l5°–20° C) which is less than ideal for most other fishes. Thus an upward shift in temperature of only a few degrees can result in a major change in the fish fauna. Such changes can be caused by diversions that reduce flow, logging that opens the forest canopy over a stream, or a variety of natural factors. For example, the stream I mentioned earlier had its waters warmed by a small impoundment behind a flood control dam. Because of the warmer water, the low-gradient section of stream below the dam contained not only trout but large populations of suckers and minnows as well. About two kilometers downstream, the gradient increased and temperatures decreased thanks to the inflow of a couple of cold springs. This lower section contained mainly trout and sculpin. This arrangement was the exact opposite of the distribution pattern found in most streams and demonstrates the importance of environmental quality in maintaining trout populations.

Another major factor limiting fish abundance is **water chemistry**. Trout, for example, are active fish and require water nearly saturated in

Figure 7-3. A fairly natural section of Rush Creek in California. Note the deep pool and the riparian vegetation that provide cover for fish.

oxygen to survive. Thus sewage pollution, which uses up the oxygen in the water, quickly eliminates trout. Similarly, most fishes cannot stand very acid water; in the southeastern United States there are many beautiful clear streams that contain few fish because the water drains coal and other mines and leaches out acidic materials. Sometimes even a minor event can cause a major change in a stream. In the Great Smoky Mountains, construction of a single road crossing across a trout stream drastically reduced the trout populations because it dumped acidic rock in the stream bed.

One other factor that is particularly important in clear trout streams is **cover**. All fish need a place to hide, whether it is from each other, from kingfishers and herons, from snakes and otters, or from potential prey. The problem in defining cover is that each species and size class within a species uses different kinds of cover. A large trout may need an undercut bank in a meter or more of water, whereas a small one may need only an overhanging rock in a few centimeters of water, or even just the rippled surface fast water provides. The greater the diversity of kinds of cover in a stream, the more different kinds and sizes of fish are likely to be present. This is easy to see just by looking closely at a partially submerged bush or tree in a pool; chances are there will be schools of small fish swimming among the branches, protected from a predatory trout lurking in the deeper water below. Pools without such cover usually lack small fish.

One of the main effects of stream channelization (i.e., the practice of turning a natural stream into a ditch for rapid drainage) is to eliminate most of the cover for fishes. In the channelized sections of a California stream I studied, not only were there fewer fish than in the unchannelized sections, but the fish were much smaller. There was essentially no cover left for catchable-sized trout favored by anglers, much less an endangered sucker species that lived there as well.

PREDATORS, COMPETITORS, AND PARASITES

Cover is important to fishes because predation is probably the most frequent cause of death in natural populations, especially among smaller fishes. Trout streams have a wide array of piscivorous (fish-eating) animals associated with them. Among the smallest are some of the same aquatic insects that are important prey themselves for large trout. Stonefly larvae, dobson fly larvae, and the larvae and adults of some beetles often consume newly hatched trout. Large trout can also be voracious predators on their own young, which is presumably one of the main reasons that small trout are most abundant along the edges of streams or in small tributaries where large trout cannot go. The most piscivorous trout in streams are generally brown trout; introduction of brown trout into a stream may be followed by a decline in the abundance of native trout and other fish. Also effective as piscivores are snakes, especially aquatic garter snakes in western North America and "true" water snakes in eastern North America. These snakes are most effective at capturing fish when water levels are low, but their commonness is indicative of their general success at ambushing small fishes. Usually the most important predators on trout streams (beside humans) are birds,

Figure 7-4. *Top:* A section of Rush Creek, channelized to prevent a pasture from flooding. *Bottom:* A section channelized ten years earlier that has eroded down to bedrock and in which grazing by sheep has prevented the riparian plants from restoring themselves.

because their high energy needs make the consumption of large amounts of prey mandatory. Most common are:

1. *Kingfishers,* which take fish by diving on them from above.

2. *Mergansers,* or "fish-ducks," which may occur in flocks of ten or more cooperating birds that swim after their prey.

Total Number Caught

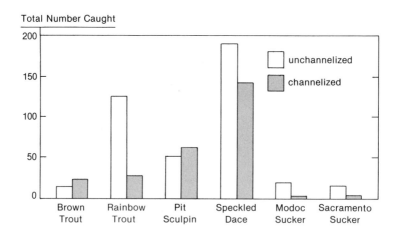

Grams Per Square Meter

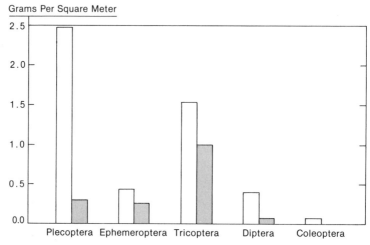

Figure 7-5. *Top:* Fish populations of channelized and unchannelized sections of Rush Creek. Not only are fish (especially trout and the endangered Modoc sucker) less abundant in the channelized sections but they are much smaller on the average. *Bottom:* Invertebrates of all types are also much less abundant in channelized sections of stream.

3. *Herons* of various species, which wade quietly in streams and spear passing fish.

A number of mammals also prey on fish, but most common are mink and otter. Otter in fact live mainly on fish and crayfish and pursue the same size fish that anglers like to catch. Their most common prey are

water snake

belted kingfisher

trout hatchling   stonefly

Figure 7-6.   Some predators commonly found on trout streams. Their effect on trout populations is surprisingly small.

suckers and other nongame fish if they are available, presumably because they are easier to catch than trout.

This catalog of predators has long been of concern to anglers who often feel that the natural predators compete with them for their prey. By this logic, elimination of the predators should improve the catch of the anglers. However, predator control programs are rarely effective. One reason is that most such programs have been incomplete; they may kill the birds and mammals but not the large brown trout, for example. More important, such programs often ignore the fact that even if the number of fish can be increased in a stream, it doesn't mean that there is enough food or space to support the additional fish. The additional fish may be displaced downstream by hungry competitors of the same species, until they reach a point where the predators are not controlled or until they reach an unfavorable environment and die of other causes. Nevertheless predator control programs have been successful on a number of salmon streams in greatly increasing the number of young salmon going out to sea. Unfortunately, there is little evidence that this increases the number of adult salmon that return a few years later because, as explained in the previous chapter, decreased predation in the stream is compensated for by increased predation in the ocean.

Competition, like predation, is often thought to limit trout and salmon populations. Nongame fishes, especially, are often accused of depressing salmonid populations although hard evidence is scanty. Usually, the accusations come following the decline of trout populations accompanied by an increase in sucker populations. It is assumed that the latter causes the former when usually it is a change in environmental conditions that is at fault, such as logging of a watershed. Such factors can warm up the water, silt in deep pools needed for cover, or cause an increase in the severity of floods. The changed conditions may favor nongame fishes but not trout. When degraded watersheds are restored, the trout often return and the nongame fishes decline. When underwater observations are made of direct interactions among fishes, you can occasionally observe a trout chase another fish, but rarely will a trout be chased or disturbed by any fish other than another trout. In fact the most severe competition that takes place in trout streams is among coexisting trout and salmon species, especially where one species is a recent introduction into the stream. Brown trout, for example, can displace native brook trout from the best holding places in a stream. The displaced brook trout wind up in less protected areas where food is harder to obtain and where they are more likely to be caught by anglers or other predators. The result is that brook trout streams often become brown trout streams in a few years. In the West, similar events have taken place in cutthroat trout and golden trout streams.

The reason that trout interactions can be so severe is that trout are naturally very aggressive (which makes them good game fish) and all trout and salmon species seem to use the same basic set of signals in their behavior. Thus when a territory-holding brown trout flares its gill covers at a brook trout, the brook trout will perceive that as a threat and probably flee, especially if it is smaller than the brown trout. In a stream where only brook trout live, these behavior patterns serve the species very well, because they force individuals to spread out and make maximum use of the limited food and space available.

An unpredictable factor that may affect the outcome of competition or predation among fishes is the presence of **disease** or **parasites** in a fish. Virtually all fish carry some parasites with them, such as nematode worms in the gut or anchor "worms" hanging from the sides, but rarely is a seriously ill or extremely heavily parasitized fish observed. Probably most commonly observed are fish that are unwary and sluggish, with a white, furry-looking fungus called *Saprolignia* growing from a wound or injury. Such sick or injured fish are quickly seen by predators because of their abnormal behavior and so rarely last long. Fishes, like humans, develop resistance to disease organisms that are naturally present in the environment and are rarely affected by them unless the environment changes dramatically so the fish are weakened. In many trout streams, the introduction of hatchery trout has brought in new diseases, to the detriment of native wild trout. On the other side of this situation, the low survival rate of planted trout in some drainages is apparently related to their inability to survive the onslaught of native diseases.

PROJECTS

1. Aquatic insects are fascinating and easy to collect using a kick screen. A kick screen is simply a piece of standard window screen roughly 75 centimeters square tacked between two pieces of 1.5 centimeters or more dowels so one edge can be flush with the bottom. The screen is held in place in flowing water while the rocks above it are moved about in the water. The current will flush the insects onto the screen, from which they can be picked off with tweezers or washed off with a water bottle. They are easiest to observe if washed into a white pan or plate. The general types can be identified by using guides listed in chapter 15. If necessary, insects can be preserved in small jars of dilute alcohol. See if you can gain some idea of how each insect makes a living by examining it closely, especially the mouth parts, with a hand lens. Examine different habitats for different types of insects. Compare the insects you find with what you find in trout stomachs.

2. Next time you are fishing for trout, find out what they are eating by looking at their stomach contents. If you are keeping the trout to eat, cut open the stomach and wash out the contents in a small amount of water, into a small pan or net. Most of the insects will be whole and possible to identify. If you do not wish to kill the trout, a simple stomach pump can be made from a spray bottle with a squeeze handle, a 10- to 15-centimeter-long piece of plastic tubing, the kind used for aquarium air supplies, and an eyedropper. Remove the fine spray nozzle and insert the plastic tubing in its place; fit the eyedropper on the end of the tubing. The bottle can then be filled with water and the eyedrop-

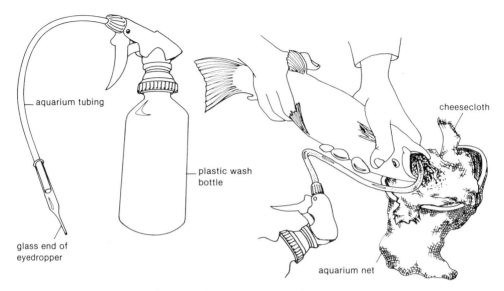

aquarium tubing

cheesecloth

plastic wash
bottle

glass end of
eyedropper

aquarium net

Figure 7-7.   A simple stomach pump to use for finding out what trout are eating. When holding the fish, be sure that your hands are wet or that you are
wearing wet gloves.

per inserted part way down the gut of the trout. The stomach contents
can then be flushed into a fine-mesh aquarium net. While handling the
trout, be sure to keep it and your hands wet, preferably by using a pair
of wet gloves. Is the trout eating mainly terrestrial or aquatic organisms?
Count the number of different food items in the stomachs of a number
of trout; does each trout seem to specialize in certain types of insects?
Do different trout specialize in different insects? If you are fishing in an
area where hatchery trout are planted on top of wild trout populations,
compare the feeding habits of the two types to see if you can find a clue
as to why hatchery trout rarely survive more than a week or two in the
wild, even if not caught. Hatchery trout can usually be recognized by
their rounded and frayed fins and their chunky bodies and small heads.

   3. With a mask, snorkel, and wetsuit (unless you enjoy cold water),
spend some time observing trout and nongame fish in a stream. Note
that trout are generally much easier to observe than nongame fish and
can usually be observed feeding in fast water. Are the different species
of fish found in the same place? Are they feeding in the same manner?
Crawl up into the fast water of a riffle and observe the insects and sculpins, if present. Sculpins are often best observed initially by carefully
turning over rocks under which they are hiding.

CHAPTER VIII

Warmwater Streams

For many people, trout streams form the image of the ideal stream, but in reality there are many more miles of warmwater streams, which are often much more accessible to the average citizen. For the naturalist, small- to medium-sized warmwater streams present one of the most interesting aquatic habitats to study because they typically contain many species and individuals of fish (and other creatures). The stream closest to my own home in Davis, California, is Putah Creek, which has a surprisingly rich fish fauna, despite years of neglect and abuse. It is a natural attraction for local children who do not mind catching small sunfish, as well as to naturalists who come to catch glimpses of beaver, wood ducks, and brilliant dragonflies. Every time I walk along the creek, I delight in something new, from busy colonies of spawning sunfish, to scats and tracks from a passing otter, to hawks flying overhead, to floating puffs of cottonwood seed.

Unfortunately, streams like Putah Creek are among the most abused aquatic habitats; they have long been convenient dumping grounds for sewage and other pollutants. Their tendency to flood on occasion means most of them have been subjected to channelization, bank alterations, damming, and other means of bringing them "under control." Until recently, warmwater streams were abused without much protest, because relatively few of them support the intense fisheries found on trout streams. The most valued fishes are smallmouth bass and pickerel, but even they are abundant in only a small percentage of the streams; cat-

Figure 8-1.   The Susquehanna River, as it flows past Three Mile Island nuclear plant at Goldsboro, Pennsylvania, is a warmwater stream rich in fishes and rich in problems. From a fish's perspective, radiation released from the plant is a miniscule problem compared to other pollutants in the river.

fish, sauger, largemouth bass, and other game fishes are important mainly in the large rivers. Often sport fisheries exist in the smaller streams for chubs, suckers, carp, and other "nongame" fishes but these fisheries rarely have vocal or influential supporters. As a result, the streams have suffered from neglect by fisheries managers and from lack of study by biologists. Fortunately, this is changing and many biologists like myself have found their local warmwater streams fascinating places to study. The scientific literature on these streams is growing rapidly as a consequence. In this chapter, only the small- to medium-sized streams will be discussed because they are the most accessible for study and observation.

THE NATURE OF WARMWATER
STREAMS

Temperate warmwater streams are typically found at low elevations. Rapids and waterfalls are present, but much more common are long pools and runs (glides) and shallow gravelly riffles. During the summer months, water temperatures are in excess of 23°–25° C for extended periods of time, so coldwater fishes, such as trout and sculpins, are excluded. Otherwise the streams are highly variable. They can range from extremely muddy to crystal clear, flowing through limestone canyons or meandering through flat valleys. Their bottoms range from mud to bedrock, but they characteristically have extensive cover created by fallen trees and other riparian debris. If the headwaters of these streams are at high elevations, they may start out as trout streams, gradually changing to warmwater streams at lower elevations; the fish fauna changes as well, in zones as discussed in the previous chapter. Often, however, the headwaters are warm as well and are often intermittent in flow. In these situations, the fish fauna gradually becomes more diverse in a downstream direction. In moister areas, the headwater fishes will be those that can live in small, moderately fast streams, such as creek chub and blacknose dace (eastern United States), whereas in drier areas, the headwater species are likely to be those that can survive in pools left behind as the stream dries up or can colonize areas that are only seasonally available (green sunfish, fathead minnow).

In eastern North America, the number of species found in warmwater streams is often surprisingly large. Species lists of twenty to fifty are common, although the relative abundances of the species may change considerably from year to year. The number of species increases from headwater to mouth, because as stream size increases, the variety of habitat also increases. The lower reaches of streams are often used on a temporary basis for spawning by fishes from larger rivers but the number of species typically found in a river may be less than that found in the smaller streams that feed it. In the smaller streams the dominant fishes are minnows, especially shiners of the genus *Notropis*, darters (small, perch-like bottom fishes), sunfishes, suckers, and madtoms (small catfishes). The number of species is high because water is constantly available (the streams rarely dry up completely even during major droughts) and because the streams are part of the great Mississippi–Missouri River drainage, or had connections to it at one time. This drainage is not only enormous, connecting the waterways of much of

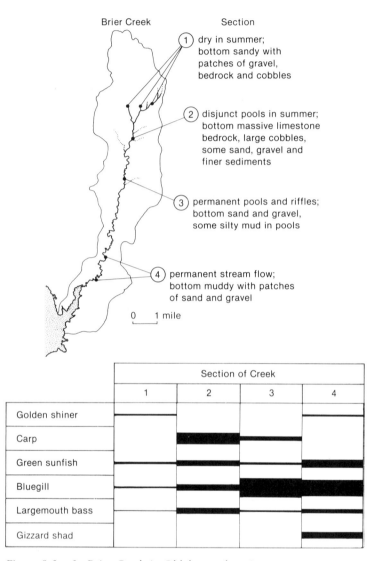

Brier Creek — Section

1 dry in summer; bottom sandy with patches of gravel, bedrock and cobbles

2 disjunct pools in summer; bottom massive limestone bedrock, large cobbles, some sand, gravel and finer sediments

3 permanent pools and riffles; bottom sand and gravel, some silty mud in pools

4 permanent stream flow; bottom muddy with patches of sand and gravel

0   1 mile

	Section of Creek			
	1	2	3	4
Golden shiner	▬			▬
Carp		▬	▬	
Green sunfish	▬	▬		▬
Bluegill	▬	▬	▬	▬
Largemouth bass	▬	▬	▬	▬
Gizzard shad				▬

Figure 8-2. In Brier Creek in Oklahoma there is a strong association between fish species and habitat type. The headwaters (1) are a very harsh environment, inhabited mainly by species that are good colonizers, whereas the creek mouth (4) is inhabited both by stream fishes and by those that live mainly in the reservoir, "Lake" Texoma, into which the creek flows.

North America east of the Rocky Mountains, but it is very old. It has thus served as a refuge for "ancient" fishes, such as gars, bowfin, paddlefish, sturgeon, mooneye, and others, and provided many opportunities for the wide dispersal of species that evolved in tributary systems.

In contrast, the streams of western North America contain many fewer species. Because the major drainages are isolated from each other by mountain ranges, the number of **endemic** species (those confined to single drainages) is high and often adjacent drainages will share few species. For example, the Sacramento–San Joaquin drainage (California) on the west side of the Sierra Nevada shares only one native species (speckled dace) with the drainages of the Great Basin on the east side. A high percentage of the species are large as adults and are long-lived. This appears to be an adaptation to surviving in environments where droughts are common and small tributaries dry up on occasion. Large fish can persist in the deep pools of the larger streams, outlasting the drought to spawn successfully during wetter years. Relatively few species can survive the rigorous annual flow regime found in most western waterways: floods in the spring from the runoff of melting snow followed by extreme low flows created by rainless summers. This pattern becomes less pronounced toward the north, disappearing in Canada, as the streams become dominated by salmon and trout. In more southern drainages, such as the Colorado River, the dominant fishes are large minnows (such as squawfish and bonytail chub) and suckers (such as flannelmouth and razorback suckers).

The timing, frequency, and size of floods are major factors affecting fishes in warmwater streams. In the western United States, the timing of high flows, in late spring, is fairly predictable. Consequently, almost all the resident stream fishes spawn in the spring. Large adults can move up into small tributaries at this time to spawn; these tributaries, if they do not dry up completely, serve as nursery grounds for the young fish, where they can be relatively free from competition and predation from adults.

Spring is also the peak time for spawning in the East, especially among the larger fishes, because flows are usually higher then as well and migrations are possible. For some species, such as carp and carp suckers, spring spawning allows fish to take advantage of flooded vegetation for protection of eggs and young. However, severe floods in the East can occur at almost any time of year, caused by such factors as winter thaws and summer thunderstorms. Such floods result in sudden increases in flows and may last only a short while but can be disasters for some populations of fish. In a Minnesota stream, for example, a major thunderstorm caused a flash flood while rainbow darters were breeding in the fast water of riffles. After the flood, Thomas Coon, who had been studying them, could not locate any individuals of this species in the stream. Two other species of darters that were not breeding at the time were not affected by the flood. Sudden floods may also wash away young fishes that depend on warm, quiet water at the stream edge for survival. Presumably because of such unpredictable floods the species of small fishes in eastern streams have different spawning times or have protracted spawning periods. An early flood that is a disaster for one species may be beneficial for another species that spawns later because the populations of a competitor (or predator) have been reduced.

FEEDING HABITS OF WARMWATER FISHES

One of the more interesting questions raised by the complex fish communities of warmwater streams is "How do all those fishes manage to live together?" One way to get at least a partial answer to that question is to look at the diets of the fishes to see how the species divide up the possible food items available in their environment. Surprisingly, the fishes generally show a great deal of overlap in diet among species; narrow specialists are few, whereas generalists are common. Even so, it is possible to divide the fishes into groups based on similar feeding habits: herbivores, omnivores, bottom invertebrate feeders, water column feeders, and piscivores.

Herbivores. Fish species that feed largely on algae and aquatic plants in streams are not many and this category is not clearly separated from the omnivore category. Both types of fishes have long, convoluted intestines and downward-oriented mouths, for grazing on the bottom. Most herbivores are minnows, although pupfishes may occupy this role in the Southwest. The native herbivores feed mainly on algae, which they may keep grazed down to low levels. In eastern streams, the principal grazer is the stoneroller, an abundant minnow, 10–15 centimeters long. Mary Power and William Matthews showed that where stonerollers were abundant in Oklahoma streams, algae was not, except in shallow water, which fish avoided in order not to be captured by herons. Algae also flourished in pools kept clear of stonerollers by piscivorous largemouth bass.

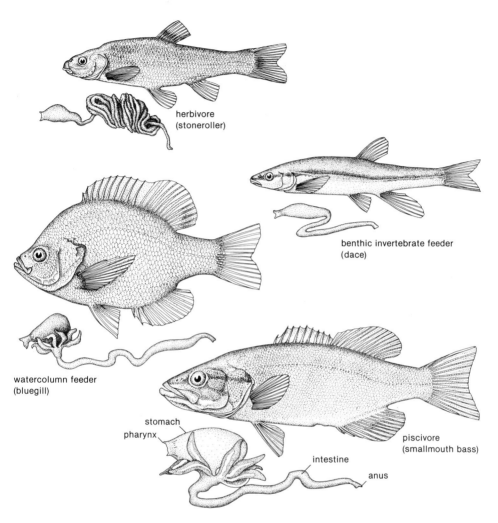

herbivore
(stoneroller)

benthic invertebrate feeder
(dace)

watercolumn feeder
(bluegill)

stomach
pharynx

intestine

anus

piscivore
(smallmouth bass)

Figure 8-3.   The diets of warmwater fishes are reflected in the anatomy of the gut. Herbivorous minnows have long, convoluted intestines; those of insect-feeding minnows are short, without a true stomach. Larger predators have an expandable stomach to handle large insects (bluegill) or fish and crayfish (smallmouth bass).

Omnivores.   These fishes have a high percentage of algae in their diets but also feed equally on everything ranging from organic ooze to aquatic insects. Most conspicuous of the omnivores are suckers which are found in large numbers in streams throughout North America. The amount of plant and animal matter in the guts of suckers varies from place to place, time to time, and among species, but all of this food is obtained by systematically moving over the bottom, finding and sucking up bits of food through their sensitive, flexible lips. Another group of omnivores are catfishes, especially bullheads, which will scoop up anything edible from algae to insects to dead fish.

Bottom invertebrate feeders.   Although omnivores often seem to dominate streams because of their large sizes, invertebrate feeders (bottom and water column) usually dominate by numbers and species. Indeed, most stream fishes are invertebrate feeders during the early stages of their life histories. Fishes that feed on bottom invertebrates are mainly small species that search among the rocks and vegetation for aquatic insects, snails, and other invertebrates. In eastern streams, this group includes the many species of darters and many small minnows such as blacknose dace, longnose dace, and suckermouth minnow. These fishes are active mainly during the day; at night the principal seekers of invertebrates are the madtoms, small secretive catfishes. In western streams, darters and madtoms are absent, so the chief benthic invertebrate feeders are small minnows like speckled dace and the juveniles of larger species like hardhead and squawfish.

Water column feeders.   Among the most available sources of food in streams are the terrestrial insects that fall into the water from the surrounding vegetation and the aquatic insects that emerge to breed and lay eggs. Aquatic insects are also available as drift, as in trout streams. There are many fishes that take advantage of the insects in the water column or at the water's surface: schools of shiners, sunfishes, small bass of various species, and others. These fishes can be observed foraging at the heads of pools and along the vegetated edges, where terrestrial insects are most likely to fall in.

During hatches of aquatic insects, the quiet waters of pools will be dappled by breaks caused by feeding fishes. Most of these fishes are actually quite opportunistic: they will feed on the bottom or among aquatic plants if a food organism presents itself. Among the shiners, it is common to observe several species of similar size and shape in one pool. Minor differences in the orientation of their mouths, however, reflect differences in where each species feeds most of the time. Thus the emerald shiner has a slightly upturned mouth and feeds more on the surface, whereas the bigmouth shiner, with a slightly downward-pointing mouth, feeds more on the bottom. Other species with straight mouths may feed more in the water column itself.

Piscivores.   Fishes that prey on their fellow vertebrates and on large invertebrates such as crayfish are the most favored game fishes in warmwater streams, but they are often surprisingly scarce. The smaller the stream, the fewer piscivores will be found. This is presumably because they require deep pools and cover to avoid their own predators, such as humans and otters. Typical examples are large and smallmouth bass, squawfish, grass pickerel, fallfish, and channel catfish.

BREEDING HABITS

Many stream fishes assume spectacular coloration while they are breeding; breeding colors of the red shiner, redbelly dace, rainbow darter, and many others rival those of tropical aquarium fishes. The minnows and suckers also are covered with breeding tubercles, small hard protuberances that are best developed in males, especially of species like the hornyhead chub and stoneroller. Many of these species build and defend nests and are easy to observe while doing so. One of the most spectacular nests is that of the hornyhead chub, a minnow that may reach 20 centimeters in length. The large males build a raised nest of pebbles that grows continuously during the spawning season and may eventually get as large as a meter long, a meter wide and 15 centimeters high. The male defends his nest against other males and egg predators (such as small hogsuckers) and entices as many local females as possible to spawn with him. The breeding tubercles function in these activities much like antlers of deer—for defense, real and ritualized, and for attracting females.

Because the loose pile of pebbles constructed by the chub seems to be an ideal spot for the incubation of eggs, other species of minnows, mainly small shiners, are attracted to it for spawning as well. Different species may be spawning simultaneously on different parts of the pile and one of the results of this is the accidental mixing of eggs and sperm from different species. The consequences of this are hybrid fishes, which have characters intermediate between the two parent species. Such hybrids are usually incapable of breeding themselves, so the parent species retain their distinct identities.

Multi-species spawning also occurs in the nests of minnows such as creek chub and stoneroller. Creek chub build shallow pits for spawning and deposit pebbles from the excavation upstream. A large male may spawn many times and constructs a new excavation each time, burying the previous one. The eventual result is a long ridge of gravel with a pit at the downstream end, guarded by the male.

Sunfishes and basses construct nests in quiet backwater areas. The nests are defended vigorously from other males of the same species and from egg predators, but, curiously, small minnows occasionally will spawn in the defended nests of largemouth bass. Most sunfishes are colonial nesters and the spawning ground may be identified not only by the concentration of fish, but by the concentration of plate-sized excavations. Other nest-building species are less conspicuous. Johnny darters, for example, spawn under carefully selected flat rocks, in "caves." The fish spawn upside down, because the eggs are attached to the roof of the cave. Similar behavior is characteristic of some minnows (e.g., fathead and bluntnose minnows) and of the small madtom catfishes.

Even more common than nest-building behavior (especially in western streams) is mass spawning behavior, which is most noticeable among large suckers. Here the fishes aggregate on a gravel riffle. The males are usually most conspicuous because they tend to distribute themselves over the riffle, waiting for females to enter from the pool downstream. When a receptive female enters, she quickly gains the attendance of two or three males and spawning occurs rapidly. Fishes (particularly males) spawning in this manner are especially vulnerable to predation, so at the end of the season females are likely to outnumber males!

PROJECTS

1. One of the real challenges in working with warmwater stream fishes is simply identifying them because there are so many similar species that live together. Learning to use a key to the fishes is vital if you wish to do this. See chapter 15 for suggestions for books useful in your area. If you decide to collect fishes from a local stream, be sure to check with your state agency in charge of fisheries to find out regulations regarding collecting gear that is legal. You will probably need a fishing license as well.

2. In early summer or late spring, spend some time exploring your local stream looking for nest depressions and ridges or for aggregations of spawning fishes. From a good vantage point on the bank quietly ob-

serve breeding activity and interactions among males and among species. Early morning is often the best time for such observations.

3. Through observations in a number of habitats and over several years, become well acquainted with the fish fauna of a local stream. Small warmwater streams often need stream guardians, individuals who are familiar with the fishes (and other animals) and so can determine if pollution, channelization, or some other unnatural disaster is affecting the health of the stream. In many areas, groups have organized to protect local streams, The Friends of "X" Creek. The most difficult problem such groups face is providing evidence that a stream is being harmed. Careful notes (especially if published in a newsletter) and photographs, taken across seasons and over years, can be a major help in this regard.

4. If you have a clear, relatively unpolluted stream nearby, spend some time snorkeling in it observing the fishes. If conditions are right (go with a friend!), try snorkeling both during the day and at night, with a light. Observe the differences between fish distributions during the day and night. In eastern streams, you will probably be able to observe madtoms foraging in the rocky riffle areas at night and observe darters during the day.

5. Set up an aquarium with local fishes and observe their behavior. A number of the species will spawn in aquaria if the right materials are provided. If you set up a permanent community tank, it is best to have a quarantine tank as well in which to keep new arrivals for several weeks to prevent the introduction of new diseases and parasites into a tank of healthy fishes. If you are really dedicated to the keeping of native fishes you can join the North American Native Fishes Association (Simon's Rock, Alford Rd., Great Barrington, Mass. 01230). If you do keep native fishes and you need to dispose of unwanted fish, either put them back into the same stream they came from or else kill them. Introducing fish into a stream to which they are not native could cause major changes in the native fish community, even the creation of endangered species, displaced by the introduced species or killed by introduced diseases.

CHAPTER IX

Lakes and Reservoirs

I had the good fortune to grow up in a house situated on a ridge between two Minnesota lakes, one large, one small. The small lake was shallow and in winter minnows would swarm to holes chopped in the ice, desperately seeking the oxygen that had disappeared from the water. One of my earlier memories is falling into the chilly water when curiosity over the swarming fish overcame caution. My mother was amused to find fish in my pockets after I managed to make it home and change out of my soaked clothes. The large lake, Lake Minnetonka, was an endless source of fish to catch with a bobber and worm: bluegill, pumpkinseed, crappie, rockbass, perch, and, as it got dark, bullheads. Minnows and silversides swarmed in the shallows, joined in spring by the vigorous splashing of spawning carp. These fish are much more visible today than they were when I was growing up because the lake has become suburbanized, resulting in installation of a sewage system and ending the lake's enrichment from leaky septic tanks.

My experiences with these lakes reflect why lakes are such popular places to live and to fish. The variety of lakes and the diversity of their fishes makes them wonderful places to study and watch fish. The changes in Lake Minnetonka, however, show how easily lakes are degraded, as well as how it is possible to restore them.

THE EPHEMERAL NATURE OF LAKES

Lakes are ephemeral in nature and their distribution reflects this. The majority occur in northern or mountainous areas that were covered with glaciers during the last ice age. The action of glaciers scours lake basins out of bedrock and deposits ridges of rock and gravel that dam streams. As they retreat they leave behind big chunks of ice that are buried in gravel deposits; as the ice melts, water-filled depressions form. Floodplains of large rivers are another place where lakes are common; the rivers wander back and forth across them, leaving fragments of old river channels behind as oxbow lakes. Such lakes are often seasonally connected to the rivers by floods. Landslides and lava flows that dam rivers are also creators of lakes, but these events are relatively uncommon.

As if to make up for the lack of lakes in many areas, public and private agencies have created thousands of reservoirs by damming streams; these are especially prominent in the arid west and in the southeastern United States. Although dam builders are fond of labeling their reservoirs as lakes, it is important to recognize that reservoirs are very different in their characteristics (and fishes) from natural lakes. Both reservoirs and lakes (but especially reservoirs) are temporary features of the landscape, as they gradually fill up with silt brought in by inflowing streams and with organic debris produced by the action of aquatic organisms.

LAKE FISHES

Given the temporary nature of most lakes, it is not surprising to find that there are few temperate fishes that are specifically adapted for lake living. Many species, such as largemouth bass, sunfish, northern pike, and chubsuckers that are abundant in lakes are also characteristic of pools and backwaters of streams. Typically, they depend on streams for their dispersal through a region and for their long-term survival as species. For most lakes, the original fish fauna is a mystery because local residents (and agencies) have seldom hesitated to introduce new species to improve fishing. This includes bait-bucket introductions from anglers who collect small fishes from nearby streams for bait and then dump the leftovers into the lake. I once spent several years studying a small lake in Minnesota, and was continually surprised by odd fishes, mostly minnows, that appeared in my samples. Most of these casual introduc-

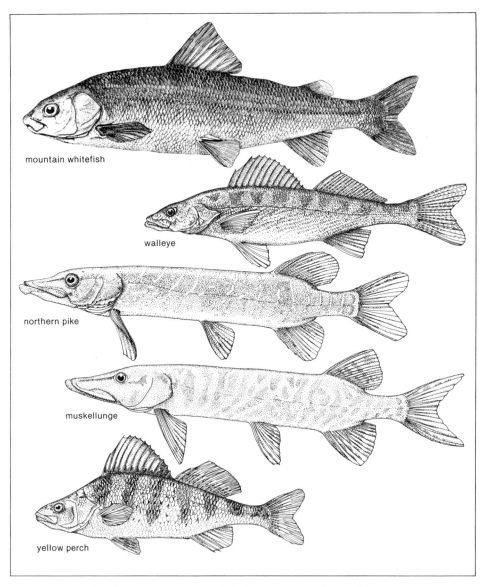

mountain whitefish

walleye

northern pike

muskellunge

yellow perch

Figure 9-1. Some common game fish that inhabit coolwater lakes.

tions never became established, but it is likely that over the years the fish fauna of the lake was gradually enriched by the few species that did manage to make it. Many isolated lakes, especially those in high mountain areas, probably had no fish in them at all until introductions were made.

TYPES OF LAKES

There are many ways to classify lakes because they are so numerous and variable. In this chapter, I will use a system commonly used by fisheries managers, recognizing that it greatly oversimplifies the picture. By this system there are four basic types of lakes that contain fish in abundance: coldwater lakes, coolwater lakes, warmwater lakes, and two-story lakes. Reservoirs can be classified in a similar fashion but will be treated separately here as a group.

Coldwater lakes.   These lakes are typical of high elevations and high latitudes. The surface waters of coldwater lakes rarely exceed 25° C and are usually considerably colder than that. Because they are located mostly in recently glaciated areas, they tend to have bedrock bottoms. Few nutrients are available to wash into these lakes, so they are clear and unproductive (**oligotrophic**). Fish populations tend to be small and individuals are slow-growing although they may reach large sizes because they often live long. The number of fish species per lake is usually low; one to five is typical. Trout and whitefish are often considered to be most typical fishes of coldwater lakes, but a number of minnows, suckers, and sculpins are adapted for life in them as well.

Lake Tahoe on the California-Nevada border is an example of a coldwater lake, although it is larger (304 square kilometers) and deeper (mean depth 313 meters) than most. Its native fish fauna consisted of a trout, a whitefish, a sculpin, a sucker, and three minnows, all of which were also found in local streams. The native cutthroat trout is now extinct, thanks to overharvest and the addition of four related species: lake trout, brown trout, rainbow trout, and kokanee salmon. The lake trout was probably the main culprit in the cutthroat extinction, because it is highly predacious and superbly adapted to coldwater lakes. It is one of the few species of trout that does not require a stream for spawning, scattering its eggs among boulders in deep water. Despite the introductions, the Lake Tahoe fish fauna seems to show a high degree of segregation by habitat and feeding habits.

Figure 9-2. *Top:* A coldwater alpine lake in the Sierra Nevada of California. *Bottom:* A coolwater lake in northern Minnesota.

Another introduction in recent years may have disrupted this picture somewhat. Following the introduction of opossum shrimp, two zooplankton species, important food organisms for small fish, nearly disappeared from the lake due to predation on them by the shrimp. As a consequence, kokanee populations have been reduced and individual fish are smaller. Large lake trout now feed to a large extent on the shrimp, suggesting that fish are less available than formerly. Similar neg-

Figure 9-2. *Top:* A riverine lake in Minnesota-Wisconsin. *Bottom:* A California reservoir.

ative effects have been noted in other large coldwater lakes following shrimp introductions.

Ironically, the introductions were made on the assumption that opossum shrimp fed mainly on non-living organic debris and would therefore fill a "vacant niche" in the lakes, providing an additional source of food for small trout. The fact that this has not worked as planned is a good indication that vacant niches by and large do not exist in nature.

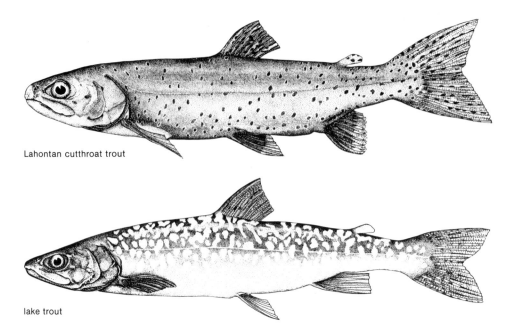

Lahontan cutthroat trout

lake trout

Figure 9-3. Introduction of the lake trout (*bottom*) into Lake Tahoe in California-Nevada was probably a major factor in the extinction of native Lahontan cutthroat trout (*top*) in the lake, although the cutthroat trout was also being overfished to supply food for logging camps. Introduction of nonnative fishes into lakes often has unforeseen consequences.

The story of Lake Tahoe's fishes demonstrates well the hazards of trying to improve a fish fauna by introducing new species.

Coolwater lakes.   These lakes are transitional between coldwater and warmwater lakes. Surface waters usually reach 20°–25° C, and the lakes are generally more productive than coldwater lakes. Trout and whitefish are often, but not always, present in low numbers, as are some of the typical fishes of warmwater lakes. Coolwater lakes are characterized by the presence of such predatory fishes as walleye, northern pike, muskellunge, yellow perch, and smallmouth bass, as well as various minnows and suckers. The predatory fishes are the mainstay of sport fisheries of the many lakes on both sides of the Canadian-U.S. border in eastern North America, such as those in the Boundary Waters Wilderness area.

Warmwater lakes.   Like warmwater streams, warmwater lakes are rich in species and show considerable year to year variation in the abundances of the fish species. They are productive and are often (but not always) turbid with blooms of algae. In clearer lakes, the production is by beds of aquatic plants. Because many of these lakes are located in farming and urban areas, they are frequently enriched by organic pollution, which tends to cause noxious-smelling blooms of bluegreen algae. The characteristic game fishes of warmwater lakes are such fishes as largemouth bass, bluegill and other sunfish, black and white crappie, bullhead catfishes, white bass, and yellow perch. Together with these fishes are numerous shiners and other minnows, darters, topminnows, and silversides. The introduced common carp is particularly successful in warmwater lakes and in shallow weedy lakes it may uproot the aquatic plants with its bottom-grubbing feeding habits, converting a clear lake into a turbid one.

Despite the number of species and their variability in abundance, the fishes of lakes tend to segregate by habitat and diet. In the rocky wave zone of shallow water, the characteristic fishes are often the same ones found in fast water in nearby streams, such as darters and dace. In sandy beach regions, large schools of shiners, silversides, and other small fishes may be most abundant, where they feed on plankton and insects and avoid the predators that live in deeper water. Sometimes these schools may consist of a mixture of species, one species oriented toward bottom browsing, another to zooplankton, and another to insects on the surface. In deeper water, one of the most complex habitats is beds of aquatic plants because they contain a mixture of plant species of different sizes and growth forms, some that may reach the surface when they flower.

Many different fishes take advantage of the food and cover provided by plant beds but this is best illustrated by examining the members of the family Centrarchidae, the black basses and sunfishes. Largemouth bass patrol above and along the edges of the beds, snapping up stray small fishes and crayfishes. Pumpkinseed (or in the South, redear sunfish) hang out among the plants in deeper water, where they pick snails off the plants and crush them with their strong, flat pharyngeal teeth. Along with pumpkinseed, but concentrating in shallower water, are bluegill, which feed mainly on insect larvae abundant on the plants. Towards the shallow ends of the bed, near logs and other cover, green sunfish and/or rock bass can be found, preying on whatever prey comes near, from small insects to fish. These two fishes are intermediate in their characteristics between bluegill and largemouth bass and seem to occupy an intermediate ecological role. The sunfishes are all fairly ag-

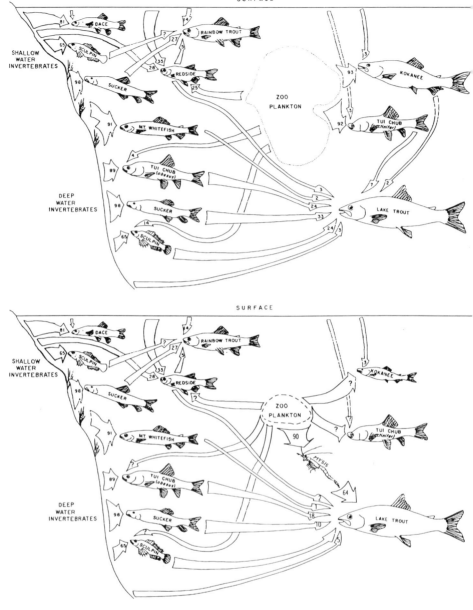

Figure 9-4. In lakes, different species typically live in different habitats and segregate by feeding habits. In Lake Tahoe, three loose assemblages of fishes can be recognized. *Top:* A shallow water (less than 1.5 m deep) assemblage dominated by Lahontan redsides, speckled dace, and rainbow trout, a deepwater bottom-oriented assemblage characterized by tui chub (*obesus* form), Paiute sculpin, Tahoe sucker, mountain whitefish, and lake trout, and a midwater assemblage characterized by tui chub (*pectinifer* form) and kokanee salmon. The numbers in the arrows indicate the percentage of the diet of each species of the item from which the arrow originates (e.g., 81% of the diet of speckled dace is shallow water invertebrates). The midwater fish community has been disrupted in recent years by the introduction of a mysid shrimp (*bottom*), which unexpectedly reduced zooplankton populations. Kokanee, which depend on zooplankton, are now less abundant in the lake and much smaller in size; presumably the same is true for the *pectinifer* tui chub as well. The lake trout now feeds mainly (64%) on shrimp and less on other fishes.

gressive species and they apparently exclude other species from their optimal feeding areas partly by chasing them away. Another fish found close by, usually in schools in or near dense cover (such as a submerged tree) is the black crappie, which moves offshore in the evening and early morning to feed on small zooplankton and small fishes. It has the odd combination of long, fine gill rakers and a large mouth, to enable it to feed on a wide variety of prey.

Two-story lakes. Many of the larger and deeper lakes stratify each summer, with a layer of warm water on top of a layer of cold water. They are separated by a *thermocline*, a narrow layer in which the temperature drops rapidly (roughly 1° C per meter). The difference between the layers can be 10°–20° C, with very little mixing occurring during summer. In more northern areas, this essentially allows a warmwater or coolwater lake to sit on top of a coldwater lake. Below the thermocline, trout may be abundant, above it, walleye, bass, and sunfish. Because the lower layer is effectively sealed off by the thermocline, it can become depleted of oxygen by late summer if there is too much organic matter decaying on the bottom. In this case, few fish are able to persist below the thermocline. During fall, the temperatures of the surface waters decrease and winds stir up the lakes; as a result the thermocline breaks down and the lake becomes uniformly cool, often down to 4° C in the winter. The warmwater fishes become less active as the water cools, but the coldwater fishes may actually show a spurt of growth at this time, as they move into shallow areas with abundant food.

An example of a two-story lake is Long Lake, Minnesota, which I studied as a graduate student (figure 9-5). It is clear, with steep sides. As a result, it has well-defined habitats that change rapidly with depth, and fish that are easy to observe using scuba gear. I recognized several zones in the lake with more or less distinct assemblages of fishes. The shallow water zone is disturbed by wave action but has overhanging bushes to provide cover from kingfishers. Large schools of mimic shiners are found swimming about here, joined by common shiners and bluntnosed minnows. Among the rocks are Iowa darters; beneath the rocks are sculpins. Next there is a brief silty-bottomed transition zone, where sunken logs are common, characterized by rockbass, green sunfish, and juvenile bluegill. This leads into the aquatic plant zone, dominated by dense beds of several species of plants. In clearings in the beds are numerous bluntnosed minnows and a few juvenile suckers, browsing on the bottom and ready to dash into cover when patrolling largemouth

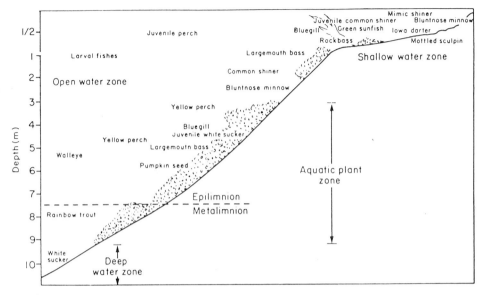

Figure 9-5. Typical locations of fishes in a Minnesota lake on a summer day. Note that there is a distinctive group of fishes inhabiting each habitat zone, but most species are associated with beds of aquatic plants.

bass pass by. Also patrolling the vegetation are a few large (about 12–15 cm) common shiners, too big to be prey for most of the bass, and yellow perch, both seeking invertebrates foolish enough to expose themselves. The perch feed on dragonfly larvae, crayfish, and other large prey, whereas the shiners seek out smaller prey. Seeking out prey on the plants themselves are bluegill (taking insects) and pumpkinseed (taking mainly snails).

The plant beds of Long Lake continue down into the thermocline, but change to beds of a single species of giant algae (*Nitella*). Few fish can be observed feeding in the algae, but in the open areas a few johnny darters and large white suckers are present on occasion. The open waters below the thermocline support rainbow trout, planted there by the local fisheries management agency. In the open waters above the thermocline, the main plankton-feeding fishes are schools of juvenile yellow perch and the microscopic larvae of other fishes. Feeding on the perch are walleye, which also move into shallow water to feed as the light dims.

The inshore movement of walleye is one of a number of changes that occurs in the lake over a 24-hour period. Large suckers may also move inshore to feed at night, whereas black crappie may move offshore in early evening to feed on zooplankton and insects in the water column. During the day, the crappie are concentrated in schools around submerged trees. Bluegill and pumpkinseed often move into shaded areas by midday, where they are relatively inactive. By being in the shade they can see predators approaching before the predators can see them. At night, after a late afternoon spent foraging, the sunfish may once again school near cover, swimming in slow circles. Other day-active fishes simply rest on the bottom; minnows lie scattered about in shallow water, whereas bass and perch rest on or beside logs and other objects. As light increases in the morning, minnow schools begin to resume formation and all the fishes begin to forage.

RESERVOIRS

Reservoirs can be classified in the same way as lakes, but a majority of them are most similar to warmwater lakes or two-story lakes in their characteristics. Here only warmwater reservoirs are discussed. Despite similarities in fish faunas to warmwater lakes, such reservoirs are very different from them. One of the biggest differences is that most reservoirs fluctuate on an annual basis; typically they are filled in the winter or spring and gradually drawn down during the summer to supply water for irrigation, urban use, or power. If a reservoir has a steady enough source of water so that it is not drawn down too far, its water level may still fluctuate several meters on an irregular basis, depending on fluctuating water demands. This means that trees and bushes rarely are able to grow down to the water's edge and that beds of aquatic plants rarely develop. Both of these factors result in lack of cover for young fishes in shallow water and a lack of diversity of habitat for adult fishes. The fluctuations also wreak havoc on the spawning of fishes such as largemouth bass and bluegill, which scoop nests out of the bottom in shallow water. These fish are likely to find their nests, with eggs and larvae, suddenly out of the water or alternately, in deep water. Shallow water habitat is usually in short supply anyway, because reservoirs typically are located in steep-sided canyons that are easy to dam.

The location of reservoirs in canyons also means that reservoirs are long and narrow, with the upper reaches being more river-like than lake-like. As a result, the fish faunas of the two ends may be fairly different. Smallmouth bass, for example, are often most abundant at the cooler,

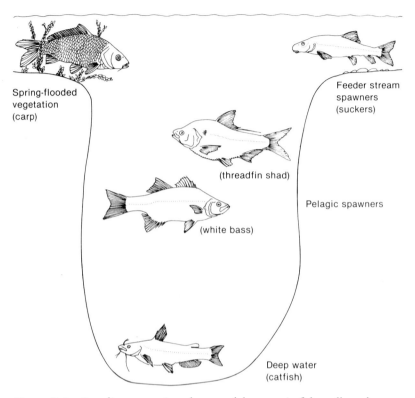

Figure 9-6.   Breeding strategies of successful reservoir fishes allow them to avoid the effects of water level fluctuations. Carp spawn on flooded plants in the early spring when water levels are high whereas suckers move up into streams to spawn. Threadfin shad and white bass have eggs and larvae that float in the water; catfishes typically nest under rocks in deep water.

flowing upstream end, whereas largemouth bass are most abundant in the warmer portions nearer the dam. In general, the fishes that do best in reservoirs are:

1.   Those that can spawn in the streams that feed them, as is the case of many suckers, or

2.   Those that can spawn in deep water, such as channel catfish, or

3.   Those that can spawn quickly on flooded annual vegetation in the spring when the reservoir is highest, as is the case for common carp, or

4.   Those that spawn in open water, with pelagic eggs and young, as is the case with white bass, gizzard shad, and threadfin shad.

Fishes with the latter two types of spawning habits are often particularly successful and may be the most abundant fishes in the reservoirs.
In contrast, many of the fishes in reservoirs are there only incidentally,
having been washed in from inlet streams. Others, mainly game fishes,
are planted on a regular basis after being raised in hatcheries, in efforts
to improve fishing. New species are continually being introduced into
reservoirs, especially in the West, in attempts to build a better fish community. In fact, the unstable nature of reservoirs means that the fish
communities will also be unstable and difficult to manage for species
of interest.

PROJECTS

1. With a minnow seine (if permitted), take samples of fish in a
sandy shallow area of a lake during both day and night. The number of
species but not necessarily the number of individuals should be higher
at night. Explain this!

2. If you fish one lake on a regular basis, keep track of your catch
according to species, time of day, depth caught, place caught, time of
year, and water temperature to see if you can detect patterns of fish
behavior, especially in relation to light and temperature.

3. By angling or seining, compare the fishes of the upstream and
downstream ends of a reservoir. In the spring, following the first draw-
down, examine shallow areas for shallow, circular depressions, former
nests of bass and other centrarchids. If information is available, compare the catch records of bass with the timing of drawdowns two or
three years previously.

4. If you have a clear lake nearby, try snorkeling in it to observe the
distribution of the fishes. Establish some transects over different habitats
you can observe on a regular basis and note the use of each habitat by
the kinds and sizes of fishes. Early in the year, locate spawning areas of
bass, bluegill, and other fishes and observe their behavior.

CHAPTER X

Ponds

Much of this book was written on a desk that overlooks a small pond in my backyard. I originally built the pond as a place to keep some of the increasingly rare native fishes that I study: Sacramento perch, Sacramento blackfish, California roach, and hitch. Thanks to neglect on my part and unauthorized additions to the fauna by family members, it now contains only a few blackfish but big schools of mosquitofish and goldfish. The pond nevertheless continues to be the center of life in my suburban back yard, full of scuds, insects, and algae as well as fish. It is also a favorite bathing spot for songbirds and is occasionally visited by lone snowy egrets or greenbacked herons that succeed in reducing the fish populations in a visit or two.

My little concrete pond is but one example of one of the most common and underappreciated aquatic habitats. Ponds vary tremendously in their characteristics, from small concrete-lined pools to small reservoirs created by constructing dams across gullies. They range from cold, spring-fed trout ponds to warm, often stagnant, farm ponds. Although they are constructed for many purposes, producing fish for sport and food is often a major justification. Fish production is also often a major disappointment because keeping a pond in a state where it can continuously produce large-sized fish requires continuous management. Regardless of a pond's ability to produce edible fishes, it is a fascinating aquatic system quite amenable to study. It can also give a naturalist considerable insight into how aquatic systems function in general. In this chapter, we will discuss only the two most common types of ponds,

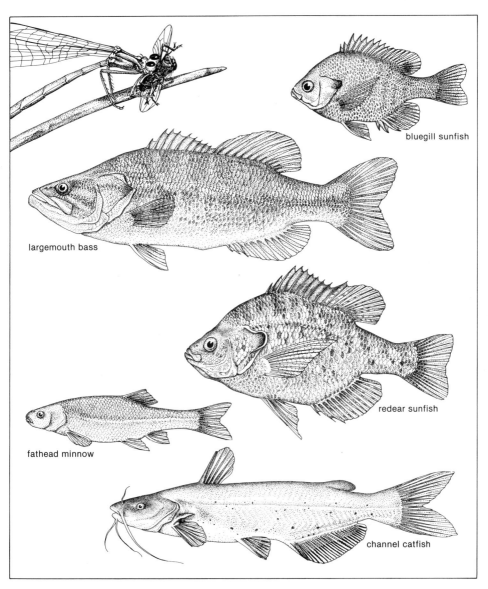

Figure 10-1. Common fishes stocked in ponds.

Figure 10-2. A large unmanaged farm pond often becomes choked with aquatic plants and the fish populations decline as a consequence.

warmwater farm ponds less than 6 acres in area and backyard ornamental fish ponds.

FARM PONDS

Farm ponds in North America typically are constructed with fishing as a secondary goal; the primary goal is storage of water, usually for livestock watering. This is changing, as awareness increases of the tremendous potential of such ponds for producing large quantities of fish, as they do in the Orient. Farm pond fishing is also increasingly promoted by fisheries management agencies as a way of satisfying the demand for fish and fishing beyond what public waters can provide. The ideal farm pond should produce continuous fishing with little management; unfortunately few such ponds exist. The typical pattern of fishing in an unmanaged warmwater pond runs something like this:

1. The pond is built and stocked with fish.
2. Fishing begins one to two years later.
3. Three to five years after construction, fishing is excellent.
4. Fishing declines after six or seven years and remains poor.

The pattern of fishing success reflects a series of natural changes that are fascinating to observe and difficult to halt. Almost as soon as a pond is filled (or even while it is filling) planktonic algae become evident as the water turns green. These algae form the base of a food web that quickly develops as zooplankton become established in the pond. The first algae-feeders to take advantage of the new pond are usually rotifers, tiny animals that are naturally found in every puddle of water. Preying on the rotifers are copepods, small invertebrates that have eggs distributed by winds blowing across dried-up ponds. In some species of copepods, even the adults can dry up, be blown around, and then rehydrate in a new location. Soon larger grazers on algae appear, mainly waterfleas. Flying insects also quickly find new ponds. Aquatic adults of predatory beetles and grazing bugs, such as waterboatmen, are soon swimming about, whereas dragonflies, damselflies, mayflies, midges, and mosquitoes drop their eggs into the water. The larvae of these insects soon appear, feeding on the algae, bottom organic matter, and each other. The larvae of mayflies and midges burrow into the bottom, whereas the larvae of dragonflies and damselflies stalk slowly over it, seeking whatever moves as prey. The small fish introduced into this situation find a rich source of food and grow rapidly to catchable size.

Usually within a year or so, aquatic plants become established in the pond, brought in with the inflowing water, as seeds passing unharmed through the digestive tracts of ducks, or on the wind. Patches of cattails and perhaps bullrushes (tules) appear along the edges, and various submerged aquatic plants start to grow in the pond itself: water lilies, with round floating leaves and attractive flowers; milfoil, with its red stems; the many varieties of pondweed (*Potamogeton*) with narrow floating leaves; coontail, with its "fuzzy" tips reaching just below the water's surface. The patches of plants become larger over the next two or three years and shallow ponds may become weed-choked. The plants are initially beneficial, because they provide homes for many insects and snails that serve as food for the fish and they provide cover for small fish. Dense growth, however, may be impenetrable for most fish and too much cover for small fish may allow them, rather than the larger fish favored by anglers, to dominate the pond. Furthermore, large masses of aquatic plants may cause organic matter to accumulate on the bottom of a pond, which may use up large amounts of oxygen during the process of decay. This creates severe shortages of oxygen in the water, especially at night, which can stress or even kill the fish.

The fish most often stocked in farm ponds are largemouth bass and bluegill sunfish. The idea behind this combination is that largemouth bass are a superb sport fish that will keep bluegill populations under control through predation, whereas bluegill are a good grazer on invertebrates that will also provide recreational fishing, especially for children. Usually, both species are planted in a new pond as fingerlings (juveniles the length of a finger!) to take advantage of the abundance of invertebrates. They grow rapidly and may even reproduce the following summer, providing abundant forage for the larger bass. Both species continue to grow rapidly and maintain their populations at modest sizes, producing the excellent fishing characteristic of many young ponds. Often channel catfish will be planted a year or so after the bass and bluegill are established, because they will feed more on the bottom than the other species and also prey on small bluegill. The catfish also grow rapidly and provide an additional dimension to the fishing but are rarely able to reproduce in ponds so must be stocked to maintain populations.

Unfortunately, this idyllic situation becomes unbalanced rather easily, usually due to a variety of factors acting simultaneously. One factor is the natural succession of events described above, where the aquatic plants make it difficult for large fish to persist, not only because their prey are harder to catch but because large fish are more likely to succumb to low oxygen levels. Another factor is selective fishing, because anglers generally catch the largest fish, especially the bass. As a consequence of reduced predation, bluegill populations may explode. At high levels, the bluegill reduce the food supply, compete with each other for the food, and become stunted, too small to be of much interest to anglers. These small bluegill may increase the problem further by harassing nesting bass and eating the eggs in bass nests. Finding ways to correct this imbalance is an interesting activity, giving an amateur naturalist an opportunity to manipulate a miniature ecosystem. The following are some possible solutions:

1. *Start again.* If a pond has a steady and inexpensive source of water and was built to be easily drained, the best way to promote good fishing is to dry it up whenever the aquatic plants become too thick. Ideally, it should be dried up long enough to kill the rootstocks as well, although for cattails and similar plants the thick roots can be removed by digging them out.

2. *Add competitors.* If overpopulation by bluegill is anticipated, redear sunfish ("shellcrackers") can be planted with them, at least in warmer regions. This sunfish will consume the same food as bluegill (although it is specialized for feeding on snails) and probably competes

with it for spawning sites. For whatever reason, stunted bluegill populations are less likely to develop in its presence. The redear itself rarely stunts. It grows as large as bluegill, but it is harder to catch so is less favored for ponds.

3. *Regulate populations artificially.* This can be done in a variety of ways. One is to simply return all large bass caught to the pond instead of eating them. If there is a local source, bass 15–20 centimeters long can be planted on an as-needed basis, to keep the number of predators high. Traps and nets can also be used to remove large numbers of small bluegill.

4. *Try other combinations.* The bass–bluegill–channel catfish combination is not sacred, just well-known. Other combinations are also possible. In California, for example, bass can be used in combination with Sacramento perch, a native sunfish. Bass have also been used in combination with other species such as golden shiners, yellow perch, and lake chubsuckers. It is even possible to have a pond without bass in it, emphasizing catfish. Channel catfish and fathead minnows make a possible combination that can produce catfish to eat and minnows to use as bait in other waters. This combination can also be used with largemouth bass because the bass are more efficient predators on the minnows than the catfish.

5. *Control aquatic weeds.* If a pond cannot be dried up, controlling aquatic weeds is likely to be difficult. Mechanical removal and herbicides result in only temporary control of the weeds. The best alternative is control through other fish species. Common carp will root up the weeds and are often available in nearby waters. Unfortunately, their rooting activity can make the pond fairly turbid. Occasionally, they will spawn successfully in such ponds and overpopulate, if bass and bluegill are not there in sufficient numbers to devour the eggs and young. Carp are actually a good food and game fish that are well suited for ponds; however, they are culturally unacceptable to most Americans. In some states, grass carp are legal and they can be used to control weeds without the attendant problems of turbidity and overpopulation, because they are unable to spawn in ponds. They have been widely promoted for weed control and are readily available. However, they should not be imported illegally into states where they are not wanted. They have considerable potential to damage natural ecosystems and can escape even from seemingly isolated ponds through such mechanisms as theft and unexpected floods.

ORNAMENTAL PONDS

The typical ornamental pond is small and cement lined, with a filtered water supply and ornamental lilies and reeds growing from submerged containers. The chief fishes in such ponds are goldfish and koi carp, which are brightly colored, hardy, and omnivorous. Such ponds can contain a considerable amount of uninvited life: dragonfly, damselfly, and midge larvae, snails, aquatic bugs and beetles, and tadpoles of frogs and toads. For a naturalist, an ornamental pond becomes increasingly interesting as the filter breaks down and the water turns green and if the goldfish and koi are replaced with local minnows and sunfishes. The ornamental pond at this point starts to resemble a convenient side pool of a local stream. If conditions are right, the fish will grow without artificial food and even spawn, if rocky substrates are provided. Phenomena such as schooling, territoriality, and predator-prey interactions can be observed. If the pond is used only seasonally, different combinations of fish can be tried in different years or fish can be transferred to aquaria indoors for the winter.

PROJECTS

1. Build a small pond in your backyard and stock it with local fishes and plants. The pond does not have to be elaborate or permanent. A 1–2 meter diameter hole with a maximum depth of 75–80 centimeters can work well, if it is lined with plastic sheeting and is located in partial shade. If the sheeting is covered with a thin layer of dirt, aquatic plants can become established. Unless you have a small pump to recirculate and aerate the water, fish densities should be kept low (but high enough to control mosquito larvae).

2. If you have access to local farm ponds, sample the fish populations of those reputed to have good fishing and those reputed to have poor fishing. Take samples of scales from bluegill from the different ponds and compare growth rates. Examine such characteristics of the ponds as area, depth, age, abundance of aquatic plants, and fishing history and see if you can come to conclusions about why the fishing is good or bad and how it could be improved.

CHAPTER XI

Estuaries

Since 1979, I have had a research project that samples fish on a monthly basis in a local estuary. The overwhelming impression of fish communities that such sampling gives is that of change: daily change with the height of the tide, seasonal change as fish move in and out in response to environmental change, and long-term change, as species become more or less abundant or as new species invade. The long-term change, unfortunately, is largely the result of human activity such as diversion of much of the inflowing fresh water for agriculture and invasions of new species, mostly carried in from abroad in the ballast water of ships. This constant change nevertheless makes estuaries fascinating places to study; something new is always happening.

Estuaries have been called "drowned river mouths" because they are bays created by the inflow of rivers to the ocean. In estuaries, large amounts of fresh water mix with salt water. This creates a very harsh environment for fish, as the rivers and oceans differ not only in salinity but in temperature, clarity, and other factors. The mixing is turbulent and uneven, because the tides move the salt water back and forth and seasonal fluctuations in river outflows vary the amount of fresh water that comes in.

The same factors that create a harsh environment also create one that is rich in nutrients. These promote the growth of algae and invertebrates, which become food for fish. The nutrients are washed into estuaries from upstream areas or result from the decay of plants growing in the marshes that ring most estuaries. Because of the mixing action, the nutrients stay in suspension and get recirculated. The overall result

delta smelt

white perch

spotted sea trout

Figure 11-1.   Some typical estuarine fishes. The delta smelt is confined to the Sacramento–San Joaquin estuary of California. White perch are common inhabitants of Atlantic coast estuaries, whereas spotted sea trout are a favorite game fish in estuaries along the Gulf of Mexico.

is an environment that is stressful to fish yet is so rich in food that it "pays" to live there to achieve rapid growth. As a result, many marine fishes, including many important in commercial and sport fisheries, spend at least part of their lives in estuaries, usually their juvenile stages. Estuaries are also favorite locations for cities, so they are heavily used by humans for many purposes, including places to dump sewage and other pollutants. Such uses generally have adverse effects on the fishes.

TYPES OF ESTUARINE FISHES

There are five broad types of fishes that live in estuaries:

1. Freshwater fishes
2. Diadromous fishes
3. True estuarine fishes
4. Nondependent marine fishes
5. Dependent marine fishes.

Usually examples of all five types can be found in one estuary, although their relative abundances vary considerably from place to place.

Freshwater fishes, for the most part, cannot survive in water more saline than 3 to 5 parts of salt per thousand parts of water (abbreviated ppt; ocean water is typically 35 ppt) and tolerant species will not live for long in water much higher than 10 ppt. This means that freshwater fishes are found mainly in the upper reaches of estuaries, close to sources of fresh water, and the amount of area they occupy depends on the amount of fresh water flowing in. Among the more salt-tolerant freshwater fishes commonly found in estuaries are common carp, white catfish, channel catfish, and mosquitofish.

Diadromous fishes are those that migrate through estuaries on their way either to fresh water or to salt water. There are two types: **anadromous** species that migrate from salt water to spawn in fresh water and **catadromous** species that migrate from fresh water to spawn in the ocean. Young of both types may spend considerable time in estuaries, taking advantage of abundant food. Salmon, lampreys, and shad are typical anadromous fishes; American eels are typical catadromous fish. Eels are especially remarkable because some of the young eels (elvers) that move in from the sea stay in the upper reaches of the estuaries and grow to adulthood there. These eels generally develop into males. Other elvers migrate long distances upstream to live and usually develop into females.

True estuarine fishes spend their entire lives in estuaries. They are few in number and most have populations that spend part of their lives outside the estuarine environment. There are few true estuarine species because of the harshness of the environment and the young geologic age of most estuaries. Examples include delta smelt (California), white perch (Atlantic coast), and spotted seatrout (Gulf Coast).

Nondependent marine fishes are fishes common in shallow oceanic waters that also exhibit some tolerance for reduced salinities. They may be abundant in estuaries but do not require them for their existence. On the Pacific coast of North America, three of the most common estuarine fishes are of this type: staghorn sculpin, starry flounder, and shiner perch.

Dependent marine fishes spend at least part of their life cycles in estuaries. Some spawn in estuaries or use them as nurseries for their young; others move into them to take advantage of abundant food. On the Atlantic coast, for example, young menhaden (family Clupeidae) and croakers (family Sciaenidae) move into estuaries a few weeks after being spawned outside of them. On the Pacific coast, there do not seem to be any species that invade estuaries only as young, except for a croaker called the totoaba. Totoaba live in the Gulf of California and use the estuary of the Colorado River as a nursery grounds. The importance of the estuary to totoaba is demonstrated by the fact it is now an endangered species; the river is nearly completely diverted so the estuarine conditions young totoaba require no longer exist most of the time.

The presence in estuaries of fishes with such widely varying uses and tolerances of estuarine conditions means that estuarine faunas change seasonally, as estuarine conditions change. When freshwater inflows are low, marine fishes may invade and predominate. During times of major floods, freshwater fishes may be found throughout an estuary. Likewise, the fishes of an estuary change from its head to its mouth, with freshwater fishes predominating at the upper end and marine fishes predominating at the lower end.

FACTORS AFFECTING DISTRIBUTION AND ABUNDANCE

Many environmental factors influence the distribution and abundance of estuarine fishes but among the most important are temperature, salinity, oxygen, and food.

Temperature is perhaps the single most important factor accounting

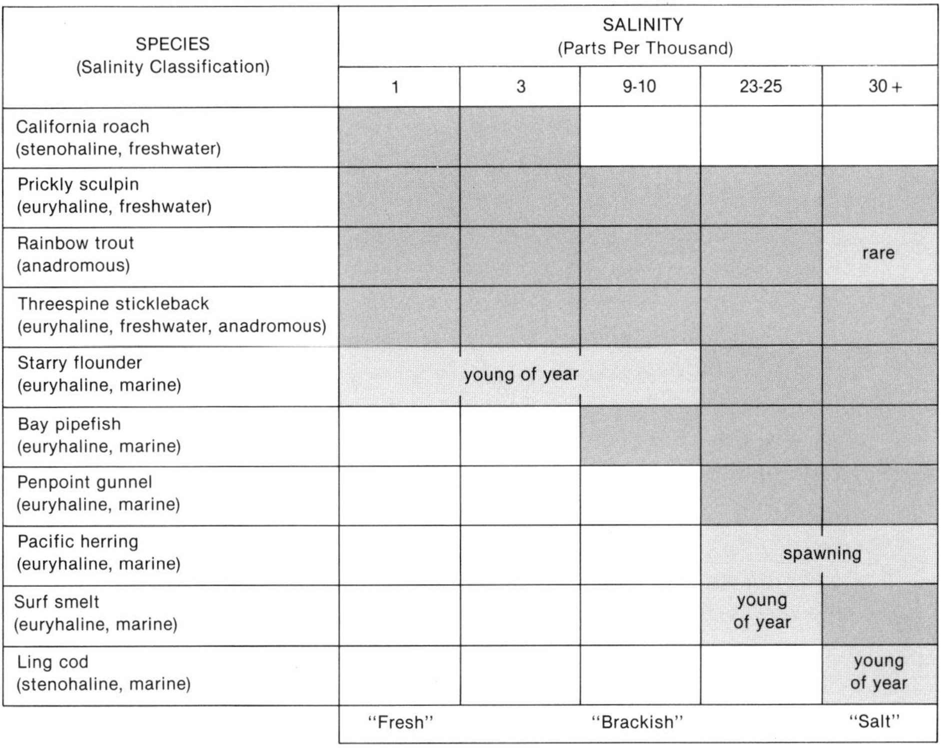

SPECIES (Salinity Classification)	SALINITY (Parts Per Thousand)				
	1	3	9-10	23-25	30 +
California roach (stenohaline, freshwater)					
Prickly sculpin (euryhaline, freshwater)					
Rainbow trout (anadromous)					rare
Threespine stickleback (euryhaline, freshwater, anadromous)					
Starry flounder (euryhaline, marine)		young of year			
Bay pipefish (euryhaline, marine)					
Penpoint gunnel (euryhaline, marine)					
Pacific herring (euryhaline, marine)				spawning	
Surf smelt (euryhaline, marine)				young of year	
Ling cod (stenohaline, marine)					young of year
	"Fresh"		"Brackish"		"Salt"

Figure 11-2.   In estuaries, the distribution of fishes is often determined by their tolerances to the salinity of the water, as shown by the distribution of fishes in the Navarro River estuary in California.

for the differences between winter and summer faunas of temperate estuaries. For example, in estuaries along the Gulf of Mexico, gulf menhaden are abundant only when temperatures reach 25° to 35° C in summer. In the Sacramento–San Joaquin estuary of California, young salmon must move out to sea by late spring to avoid lethal high temperatures. Long-term temperature trends may also affect the fishes. Declines in catches of winter flounder in Narragansett Bay in Rhode Island seem to be related to the gradual increase in water temperatures in the bay.

Estuaries are characterized by a *salinity* gradient, with salinities increasing toward the ocean. As indicated, there is a corresponding change in the fish fauna. The picture, however, is complicated by the fact that salt water is more dense than fresh water, so the inflowing fresh water will "float" on top of the salt water for a considerable distance, before being mixed by tidal and river currents. Thus in some areas, the surface waters are nearly fresh and the bottom waters are quite saline. The presence of salinity gradients is an important factor limiting the number of species in estuaries, because most marine and freshwater fishes have low tolerances for changes in salinity. One of the more successful estuarine fishes, striped bass, can survive an abrupt transfer from fresh water to full sea water, so can use the full range of estuarine conditions. Other species, such as spotted sea trout, avoid really low salinities and move to more saline portions of estuaries as freshwater inflows increase. Occasionally, sudden increases in freshwater inflow will cause mass die-offs of young fish unable to tolerate or escape from low salinities. When diversions of fresh water flowing into an estuary cause it to become more saline, marine fishes become more abundant, whereas estuarine-dependent fishes become less abundant.

Oxygen availability is usually not an important limiting factor for estuarine fishes, because the constant motion of the water keeps it well mixed and in contact with the air. Unfortunately, estuaries are all too often recipients of sewage, which consumes large amounts of oxygen in the process of decay. The result is a great decrease in the amount of oxygen available to fishes, especially bottom fishes, and a consequent decline in the numbers and kinds of fishes present. In the estuary of the Thames River, England, a cleanup of sewage and other pollutants resulted in a dramatic return of the fishes, including Atlantic salmon. The cleanup of the Thames provides an example for the rest of the world to follow.

Food is one of the main attractions of estuaries to fish because it is typically concentrated there. The zone of estuaries where the active mixing of fresh and salt water takes place is especially rich in food organisms because the physical processes concentrate them, keeping them circulating rather than being flushed out. Not surprisingly, the reproductive strategy of many estuarine fishes is to time and place their spawning so their young wind up in the mixing zone. By taking advantage of abundant food, the young can grow rapidly.

Because of the rigorous nature of the estuarine environment, relatively few species are found in an estuary at any one time. As a result, food webs are typically fairly simple. Most estuarine fishes have fairly broad diets, so they can adjust to the shifting abundances of prey items. The basis of most estuarine food webs is detritus, fine organic material coming from surrounding terrestrial or marsh plant communities. The extensive salt marshes associated with estuaries are particularly important as a source of detritus. The fine particles of detritus are gathered by filter-feeding zooplankton, shrimp, and clams which in turn become food for fish. In the Sacramento–San Joaquin estuary, the most important filter feeder is (or was) just one species of opossum shrimp which virtually every fish species feeds on when it is abundant. As this is being written, a new filter feeder, a small clam from China, is colonizing the estuary explosively. The clam has become so abundant that in some parts of the estuary it is filtering out most of the plankton. This is likely to have severe negative effects on the populations of larval fish, which also depend on plankton for food.

STRIPED BASS: AN ESTUARINE FISH OF BOTH COASTS

The striped bass is, or has been, one the most important estuarine-dependent fishes on both the Atlantic and Pacific coasts of the United States. It was originally found only on the East Coast but was introduced into California in 1879 and 1882. Only 432 fish survived in two separate trips across country on the new transcontinental railroad, but within twenty years the harvest was being measured in millions of pounds. It was imported to California because on the East Coast it had long been considered one of the finest eating fishes and a fine sport fish as well. The reason for its success and popularity is that the striped bass is a predatory fish superbly adapted for estuarine conditions. It can live in nearly the entire range of natural conditions found in its estuaries and withstand abrupt changes in both temperature and salinity. It is capable of making extensive migrations to find food and spawning grounds and will feed on a wide variety of fish and invertebrates. It is also very prolific; a large female can produce one to two million eggs in a season.

In spring, when river flows are fairly high, adult striped bass move upstream to spawn, migrating to the deep river reaches above the estuaries, and spawning when temperatures are 15°–20° C. The bass spawn in open water. Several males pursue each spawning female, their dorsal fins often breaking the surface during the chase. When the female releases some eggs, the males release clouds of milt to fertilize them. The fertilized eggs sink to the bottom but do not stay there. They are just slightly buoyant so are carried along the bottom by the current. The eggs bounce along the bottom for a couple of days until they hatch and the helpless larvae are then carried even further. If the bass have spawned in the right place, the larvae begin to feed just at the time they reach the mixing zone, where food concentrations are highest. They grow

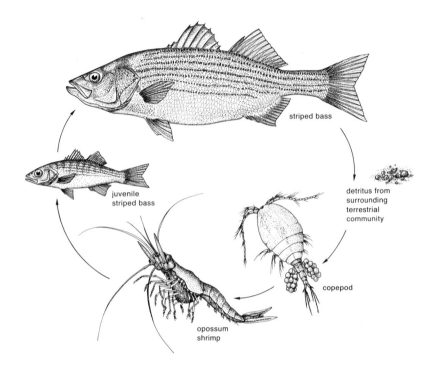

Figure 11-3.   Estuarine food chains are often simple. In the Sacramento–San Joaquin estuary, adult striped bass feed on their own young (and other fishes), which prey on opossum shrimp. The shrimp in turn feed largely on copepods, which graze on other kinds of zooplankton and on algae.

rapidly, feeding voraciously on zooplankton and small shrimp, typically becoming 3 to 5 centimeters long by the end of their first summer.

In their second year, when the bass reach a length of about 10 centimeters, their diet switches increasingly to small fish, including smaller striped bass. During the second year, on the Atlantic coast, the bass usually move out to sea. On the Pacific coast, some bass may move out to sea, but most stay in the estuaries. The East Coast populations stay close to the coast as adults, making seasonal migrations in search of optimal conditions for growth and returning every year to spawn in the estuary. They may live over thirty years and reach lengths of nearly 2 meters; the largest and oldest fish are invariably females.

On the Atlantic coast, the center of striped bass abundance is the Chesapeake Bay estuarine system, whereas on the Pacific coast it is the Sacramento–San Joaquin estuary. In both areas there have been major declines in striped bass populations in recent years. The causes of the declines center around various abuses of the estuaries by humans. In both cases, there has been poor survival of young fish combined with probable overharvest of adults, both legal and illegal. Scientists at the U.S. Environmental Protection Agency Laboratory on Chesapeake Bay speculate that the poor survival of young bass in the bay is related to high levels of organic pollutants in inflowing waters. These pollutants enrich the waters of the bay so that there are heavy blooms of planktonic algae; the blooms shade out the beds of aquatic plants that should cover much of the bottom in shallow water. These plant beds have nearly disappeared from the bay in recent years; because these beds are prime producers of the small fish (e.g, pipefish, anchovies, croakers, silversides) striped bass feed upon, the bass may have declined as a consequence.

The sewage may also have hurt the bass populations by the fact its decay consumes much of the oxygen available in the deeper water. Under normal conditions, this deep water would be a refuge for bass when water temperatures become stressful (above 20°–25° C). However, Charles C. Coutant has gathered evidence that under low oxygen conditions, the bass cannot use the deep water refuge. They are thus forced to live in the warm surface waters and, when conditions become too severe, they die.

In the Sacramento–San Joaquin estuary, the most important factors affecting striped bass abundance are factors affecting survival in the first few months of life. There appear to be a number of factors operating simultaneously, including the larvae being exported into reservoirs, canals, and fields by water diversions, being sucked up into power plants, and being killed by toxic chemicals. In both estuaries, the bass are likely to recover only if major efforts are made to reduce abuse of the estuaries and to restore their integrity as fluctuating systems. Efforts are being made to raise striped bass in hatcheries and plant them as juveniles in order to compensate for the losses they suffer as larvae, but such a solution seems to be temporary at best. Whatever is affecting the abundance of striped bass larvae is also affecting their food supply and other fishes, so dumping hatchery bass into a depleted system may be simply throwing them away.

PROJECTS

1. Find a shallow area on the edge of an estuary you can sample with a hand seine on a regular basis, if local regulations permit. Sample it at

least monthly and keep track of the numbers and kinds of fish you catch. Note the seasonal and annual changes in fishes in relation to temperature and salinity. A rough measurement of salinity can be obtained by using a hydrometer, a simple device used for measuring water density. Hydrometers are available in aquarium stores that sell fishes for salt-water aquaria or in stores that sell wine- and beer-making supplies. If you conduct a study of this sort, be sure to obtain the proper permits from the local fisheries management agency.

2. Look up old records of fish catches from your local estuary and note the faunal changes that have taken place in the past century. Find out the causes!

3. If you fish an estuary on a regular basis, keep track of your catch by season, recording temperature and salinity each time you go out. To make comparisons, keep track of how much time you spend fishing and your catch per hour of effort. Examine the stomach contents of fish you catch and compare the contents to the catch of a small, coarse-mesh plankton net dragged through the water or to the catches of a small seine.

CHAPTER XII

Between the Tides

And joy is in the throbbing tide
Whose intricate fingers beat and glide.
        —*Rupert Brooke, "The Fish"*

In June, 1963, a seaplane taxied up to a beach in Prince William Sound, Alaska, and let me off, so I could carry my gear a short distance to a research camp. I was a student assistant for a project that was studying an unusual phenomenon: the spawning of pink and chum salmon in a long reach of Olsen Creek that was periodically flooded by salt water. The flooding was the result of the astonishing tides of the region which rose and fell 3 to 4 meters in a day. The muddy tide flats around the creek, exposed at low tide, were scenes of dramatic activity, as gulls fought over salmon carcasses deposited there, bears chased live salmon in the shallow sloughs, and shorebirds probed the mud for food. The following summer, my hike up to the camp was much longer because the Great Alaska Earthquake a few months earlier had uplifted the land, creating a long reach of new spawning habitat for the salmon. The rocks along the sound suddenly had a white "bathtub" ring of barnacles that had died after finding themselves above the reach of high tides that formerly had sustained them. In 1989, many of these same rocks were coated with gooey oil from the *Exxon Valdez* oil spill.

Oil spills and earthquakes are only some of the more dramatic events that affect the intertidal zone, the area where land and sea meet. Even a short visit to any stretch of sea coast can give witness to smaller dramas: voracious bluefish may be chasing schools of herring onto rocky New England beaches, where the herring are consumed by equally voracious gulls and terns; a small green sculpin may be snapping up small shrimp in an aquarium-like tide pool; hordes of silvery grunion may

Figure 12-1.   Rocky intertidal zone near Sonora, Mexico.

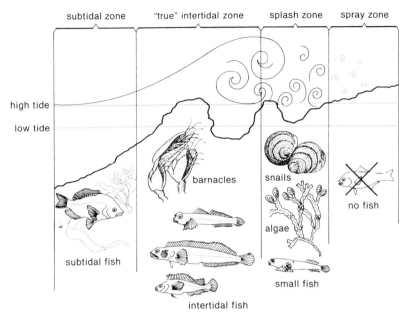

Figure 12-2.   Diagram of the rocky intertidal zone, showing the upper limits of fish.

appear late at night to spawn on a Southern California beach, their spawning carefully cued to tides and waves. At the very least, a walk along the seashore can provide evidence of past dramas: the bones of fish washed up during a storm, the cases of shark and skate eggs called "mermaid's purses," or even a dead seal with the teeth marks of a shark on it.

Clearly, the intertidal zone is a difficult place for a fish to live. Nevertheless, many fishes have special adaptations for living there because it is also rich in small invertebrates, seaweed, and other sources of food. It is consequently also a fascinating place to study fish, although to do so requires an understanding of tidal cycles. The amount of variation in tidal fluctuation varies with time of year, as the result of the complex interactions of local geography with the gravitational forces of the moon, sun, and earth. This means that it is worthwhile to obtain a regional tide table with corrections for the area of interest. Then you can make your visits coincide with the times of maximum intertidal exposure. Many types of intertidal environments can be investigated, each with its own interesting fishes. In this chapter we will consider five broad types: (1) rocky intertidal, (2) beaches, (3) mud flats, (4) salt marshes, and (5) mangrove swamps.

ROCKY INTERTIDAL FISHES

Rocky coastlines have always fascinated naturalists because of the astonishing variety of organisms that live there, thriving in a world dominated by crashing surf. The accessibility of the rocky intertidal combined with its diversity of life have made it a favorite spot for ecological studies. These studies have led to many advances in our understanding of how ecological systems work in general. Most of the pioneering work has been done with invertebrates and seaweeds, but intertidal fishes are receiving increasing attention.

One of the more striking aspects of the rocky intertidal region is the change in numbers and kinds of attached invertebrates and algae as you proceed from the uppermost reaches of the waves to areas uncovered only by the lowest tides. This change is usually described as a series of life zones. The uppermost is the **spray zone**, which is influenced by spray from waves and ocean mists but is essentially a terrestrial system. There are no fish here, except dead ones dropped by gulls. Below this is the **splash zone**, which is watered with splash from waves and may be inundated during storms. Snails (*Littorina*) are found on the rocks here, along with a few encrusting algae. Tide pools in this zone will occasionally contain small fish, washed in as larvae by high waves, but they can seldom survive the stagnation of the pools that takes place during long periods of hot calm weather. The next zone is the **true intertidal zone**, in which the upper reaches, marked by barnacles, are normally covered by high tides and the lower reaches are normally uncovered by low tides. This is the area inhabited by most of the fishes discussed in this section. Below this is a **transition zone** that is uncovered only during extreme low tides. It contains mainly the same fishes that are found among the rocks in the zone below it, the **subtidal zone**.

A wide variety of fishes use the intertidal region. They can be divided up into four basic types, depending on how they use it:

1. True residents
2. Partial residents
3. Tidal visitors
4. Seasonal visitors

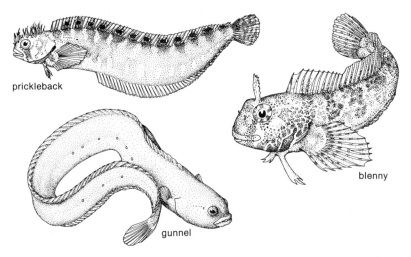

prickleback

blenny

gunnel

Figure 12-3.   Some common types of small fishes seen in tide pools; all are well adapted for hiding under or between rocks.

True residents are fishes found in the region all year around which complete their entire life cycle there; they are the fishes most often observed in tide pools. They are mostly small fishes with many specializations for living in an environment subjected to strong currents and varying salinities and temperatures, as tide pools become diluted with rain or concentrated by evaporation. Typical true residents are sculpins (family Cottidae), blennies (Blenniidae), clingfishes (Gobiesocidae), gobies (Gobiidae), pricklebacks (Stichaeidae), and gunnels (Pholidae). The sculpins found in tide pools look much like the ones found in streams because the wide, flattened head, large pectoral fins, and smooth, tapered body are good adaptations for staying on the bottom when strong currents are present. Clingfishes resemble sculpins to a certain extent, but have converted their pelvic fins into a sucker, which helps them cling to sides of rocks and other objects in strong currents. Gobies possess a pelvic sucking disc but they are not found in such fast-moving areas as clingfish; instead they use their elongate body to squeeze into cracks and crevices or for crawling under rocks. The blennies, gunnels, and pricklebacks, however, are best adapted for crevice dwelling, as their bodies are long, smooth, and eel-like, and many species have their fins reduced in size. Fishes like these come out to forage when the water is calm (high or low tide) and ambush invertebrates washed past by tidal currents. Some species, such as the monkeyface prickleback of the West Coast, browse mainly on the abundant algae in the intertidal zone.

Partial residents are fishes that are generally found in the intertidal zone, but primarily inhabit the subtidal zone. Most of these fishes are juveniles that live in the intertidal zone to avoid predators and seek out the abundant food. They typically belong to the same families as the true residents although some are juveniles of deep-bodied fishes, such as rockfish (Scorpaenidae) and surfperches (Embiotocidae). **Tidal visitors** are the common subtidal fishes that move into the intertidal zone to feed as the tide comes in. These are the fishes most often sought by anglers and include rockfish, surfperches, striped bass, and porgy. **Seasonal visitors** are tidal visitors that use intertidal areas only at certain times of year for spawning or feeding or are juvenile fishes that inhabit tide pools on a seasonal basis.

True residents and partial residents generally dominate the intertidal fish fauna, and their distribution patterns reflect the gradients of environmental severity that exist in the intertidal zone. The upper intertidal contains only a few species of small fishes that can tolerate fluctuating salinities or can survive being out of the water for short periods of time, in damp crevices or under mats of algae. At least one species of upper intertidal sculpin on the Pacific Coast has even evolved the ability to breathe air. The number of fish species increases in the lower intertidal, reflecting increased submergence time, and these species show a certain amount of segregation in both habitat and diet. For example, on the Pacific Coast, the fluffy sculpin shows strong preferences for plant cover present in deep tide pools, whereas the closely related tidepool sculpin prefers shallow water. The fluffy sculpin is typical of many intertidal fishes in that its color patterns match its background. Sculpins found on green eelgrass are bright green, whereas those found on red algae are maroon.

Because the intertidal environment is so turbulent and subject to periodic disasters ranging from sudden temperature changes to logs crashing through areas, intertidal fishes have a number of ways of keeping tide pools populated. Most have free-swimming larvae that can colonize both old and new areas, replacing adults lost through predation and other factors. Many adult intertidal fishes have strong homing abilities, so if they are forced to leave their favorite spot, they can return to it quickly. Once there, they may defend their territories against other fishes and thereby keep local food and cover for themselves.

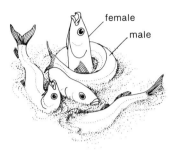

female

male

Figure 12-4. Grunion spawn on beaches during high spring tides. Here, the female has burrowed into the sand to deposit her eggs, while the male is wrapped around her in order to fertilize them. The larval grunion will emerge from the sand at the next high tide.

BEACHES AND FISHES

The surf along beaches is a high-energy environment, not only because of the immense amount of energy represented by the power of the waves but because fishes living in the surf have to expend a great deal of energy to stay there. Surprisingly, many fishes live in this surf zone, feeding on the abundant (and probably confused) invertebrates that are found there and providing sport to surf-casting anglers. Some species, such as mullet (family Mugilidae) may actually migrate through the surf, following the coast, perhaps as a way to avoid large predators. A few species spawn on beaches, the most famous being California grunion because it spawns on beaches also favored by the inhabitants of southern California. People converge on the beaches in large numbers to watch the spectacle and to collect the fish to eat. Grunion spawn at predictable times, leaving the surf for the beaches on spring nights, on high tides following a new or full moon. They are washed in by the waves and the females burrow tail first into the sand, lay their eggs, and have them fertilized by males that entwine around the females. The eggs are timed to hatch on the next series of high tides when the larvae emerge and are washed out to sea.

Other human-fish encounters on the beach are usually accidental. The leathery egg cases of some sharks and skates wash up on shore and are prized as "mermaid's purses." Occasionally, dying ribbonfish (Trachipteridae) or cutlassfish (Trichiuridae), 2 to 5 meters long, are washed up on shore and reported as sea serpents in the newspapers. Both these

Figure 12-5. The longjaw mudsucker lives on tide flats and can breathe air when necessary to survive periods of low tide.

fishes are extremely flat and elongate. Cutlassfish, in addition, have a formidable set of teeth, guaranteed to excite the imagination. On the coast of New England, silver hake often chase small fishes through the surf. So vigorous are hake in pursuit of their prey that they occasionally become stranded on beaches. Because this usually occurs in fall and winter, people who go to the beaches in search of the hake call them "frost fish."

MUD FLATS

Except for dedicated clam diggers, people rarely venture out on to the muddy flats that are associated with bays and other protected areas. From a distance, at low tide, they appear flat and barren; up close, they teem with life. Most of the organisms are invertebrates that live in burrows. A few small fishes, mainly gobies, live in the burrows as well, with or without the consent of the shrimp or worms that construct them. The blind goby of Baja California is so well adapted to the life of a burrow dweller that it resembles a cave fish, with pale skin and no eyes. Most fishes found on tide flats at low tide are confined to the narrow drainage channels; these are usually the young of common bay fishes that can tolerate fluctuating temperatures and salinities. In southern California, the longjaw mudsucker is superbly adapted for living in shallow inter-tidal puddles. Not only can it breathe air if it has to, but if conditions become too severe in one puddle, it can "walk," using its pectoral fins, to a more favorable place. The mudskippers (Gobiidae) of the tropics have similar adaptations and are so adept at "walking" they can climb up exposed tree roots.

SALT MARSHES

Mud flats eventually are colonized by salt-tolerant plants and become salt marshes. Because this process is a slow one, salt marshes in North America are extensive in the geologically quiescent Atlantic and Gulf

Figure 12-6. An Atlantic coast salt marsh and one of its most abundant fishes, the mummichog. This small (4 to 6 cm) fish feeds on the abundant invertebrates that live on the marsh vegetation and is in turn a major food for predatory birds and fishes.

coasts and relatively uncommon in the steep, geologically active Pacific coast. Salt marshes are one of the most productive environments known despite the fact they are subject to the same fluctuations in temperature, salinity, and tidal levels as all intertidal environments. They support relatively few species of fish (less than fifteen) at any given time but fish numbers may be large.

The most abundant fishes are usually **true residents**, mainly killi-fishes. These fishes have incredible abilities to survive adverse conditions, living in water ranging from fresh to several times the salinity of sea water and at temperatures ranging from near-freezing to 35 or more degrees C. The common mummichog if left behind by the tide will burrow into the mud to wait for the tide to return. When the tide does come in, these fishes will penetrate far into the marshes, in search of dense concentrations of food organisms. Often as abundant as the true residents are **partial residents**, especially schools of slender, flashy silversides. These fishes are present mainly as juveniles, although some species may be present all year.

Tidal visitors move into salt marshes to feed as the tide flows in, to take advantage of the abundant small fishes and invertebrates. Examples are croakers, flounders, and halfbeaks.

Seasonal visitors move into salt marshes at certain times of year for

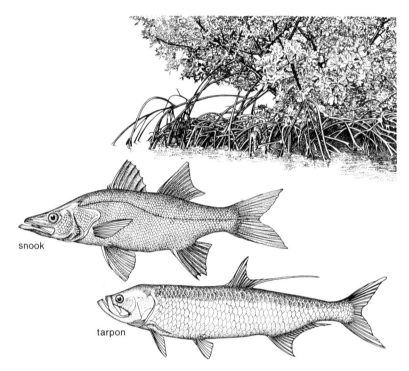

Figure 12-7.   Mangroves line the swampy coasts of much of the tropics but are confined to only a few areas in North America. Nevertheless, two important sport fishes, snook and tarpon, depend on mangroves as nursery areas for their young.

spawning, nursery grounds, or refuge from predators. On both coasts, the classic seasonal spawners are stickleback, which build nests in vegetation. In Gulf Coast salt marshes, juvenile drum, anchovy, mullet, and mojarra (Gerreidae) spend a few months of the summer in the marshes taking advantage of the abundant food and warm temperatures to increase their rate of growth. Many of the fishes that depend on salt marshes as nursery areas are also species that are harvested commercially.

MANGROVE SWAMPS

Like salt marshes, mangrove swamps are extremely productive environments that serve as nursery areas for many marine fishes. Mangroves are shrubby trees that grow in lowland areas throughout the tropics, including parts of Florida. They grow in shallow water that ranges from being more saline than seawater to being fresh. Although streams and tidal channels may meander through the mangroves, much of the backwater is stagnant, home mainly to killifishes and a few other tolerant forms. Where the water is less stagnant, juvenile fishes take advantage of the cover provided by the emergent mangrove roots and of the invertebrates that feed on the detritus from decaying mangrove leaves. In Florida, some of the game fishes that depend on mangroves in this way are tarpon, ladyfish, snook, and grey snapper.

PROJECTS

1. Collecting fishes in the rocky intertidal is generally restricted, so capture them (with a small aquarium dip net) only if you intend to return them to the pool in which you captured them. Examine the fishes you capture and look at their adaptations for living in the rocky intertidal. Many of the more cryptic species can be found underneath rocks and shells. Replace any rocks you turn over to the same position they were in before.

2. Spend some time watching one tide pool closely. If you remain still you will see a surprising amount of activity, including that of fishes. Do not get so engrossed you forget about the incoming tide and "sneaker" waves.

3. Take up poke-pole fishing as a way of discovering the larger fishes that inhabit the rocky intertidal. A poke-pole is a long bamboo pole with 15–25 centimeters of monofilament line and a hook at the end. Bait the hook with shrimp (etc.) and stick the end of the pole and line in suitable pools and crevices.

4. If you have access to a salt marsh or mangrove swamp, try setting standard minnow traps to catch killifish and other fishes or find an area that can be seined easily. Sample on a monthly basis and note the changes in species.

5. Capture some of the killifishes from a coastal environment and keep them in an aquarium to observe behavior.

6. If burrow-dwelling fishes are known to occur on tide flats of your area, try digging some out at low tide. If possible locate the invertebrate that built the burrow as well. See if some types of invertebrate burrows are favored by the fishes over other types.

7. If you are a fan of the rigorous sport of surf fishing, keep a record of the species you catch and the conditions under which you caught them, including size of waves, water temperature, tides, and the timing of the most recent storm. Is there any relationship between catch and various environmental variables?

CHAPTER XIII

The Continental Shelf and Beyond

The oceans are the planet's last great living wilderness, man's only remaining frontier on earth, and perhaps his last chance to prove himself a rational species.

— *J. L. Culliney*, The Forests of the Sea*

Because I get seasick easily, I can understand why the oceans are still a frontier. There have been times that I have concluded that the only way to prove myself part of a rational species is to get off a heaving boat as quickly as possible. Nonetheless, a trawl full of brightly colored and oddly shaped fish *usually* makes an ocean expedition worthwhile. It is astonishing how little we know about most of the fish the trawl brings in or, for that matter, what goes on in the oceans in general, even on the shallow continental shelves that ring the continents. Because of our lack of knowledge, pollution, overfishing, and other human activities are allowed to damage the marine environment. Fortunately, most coastal systems are fairly resilient and can restore themselves if a perturbation is removed or alleviated. There are also many areas that have not yet been severely altered by human activities. Both disturbed and undisturbed environments can be studied by the amateur naturalist and angler, but the difficulty and danger increases considerably with distance from shore and with depth. Because of these problems, this chapter will provide only a very brief introduction to various environments of the continental shelf in temperate areas: rocky bottoms, kelp forests, soft bottoms, sea grass flats, and open water.

*Published in 1979 (New York: Anchor Books).

Figure 13-1.   Fishes of a California kelp forest. The existence of the kelp forest depends in good part on the fish and sea otters, and the fish and sea otters benefit from the food and cover the kelp forest provides.

ROCKY BOTTOMS

In temperate areas, rocky reefs support rich fish faunas, but they are less appreciated than the faunas of tropical reefs because the coldness and comparative low clarity of the water limits accessibility to divers. Fishermen, however, have always appreciated the good fishing provided by rocky areas and have even gone to considerable trouble and expense to create artificial reefs in areas dominated by flat bottoms. These artificial reefs have been created from a variety of materials, such as old auto and bus bodies, ships deliberately sunk, concrete blocks and structures, and old tires strapped together. These reefs attempt to simulate natural habitats that provide hiding places for many species of fish, as well as rich sources of food in the form of attached invertebrates and algae.

Along the coasts, rocky areas extend up into the intertidal zone and many of the small fishes are the same types (and often the same species) that are found in the intertidal zone. Fishes such as sculpins, gunnels, pricklebacks, and blennies occupy cracks and crannies. In deeper water they are joined or replaced by such fishes as eelpouts (Zoarchidae) and wolfeels (Anarhichadidae). Roaming just above the surface or sheltering in caves or under overhangs are more active fishes, such as rockfishes (Scorpaenidae), surfperches (Embioticidae), wrasses (Labridae), porgies (Sparidae), drums (Sciaenidae), sea basses (Serranidae), and cods (Gadidae). These fishes may often occur in dense concentrations, preying on the invertebrates that also thrive in rocky areas, or rising into the water column to feed on plankton or passing schools of pelagic fishes.

In general, the diversity of fishes on rocky reefs is much higher off the west coast of North America than off the East Coast. For example, two important families of rocky reef fishes, the surfperches and the greenlings (Hexagrammidae), are largely confined to the West Coast. Studies by Alfred Ebling and his students on reefs off the coast near the University of California, Santa Barbara, have shown that the different species of surfperches on the reefs avoid competition by feeding on different kinds of invertebrates and by feeding in different ways. Some species, for example, carefully pick their prey from the bottom whereas others "winnow" by taking mouthfuls of the algal turf that covers some reefs, filtering out and swallowing the desirable organisms, and spitting out the rest.

An even more striking difference between the coasts is in the diversity of the rockfishes (*Sebastes*). These prickly bottom fishes are caught in large numbers in both sport and commercial fisheries and are frequently sold in markets as "red snapper." On the East Coast there are only two species, the abundant common redfish and the deepwater redfish. On the West Coast, over 60 species of rockfish have been described so far. They come in many different sizes and colors but otherwise have the typical rockfish characteristics such as large thorny head, large eyes and mouth, large fan-like pectoral fins, and short, streamlined body. They typically spend much of their time resting on or close to the bottom, but often feed in the water column. Although the few comparative studies of rockfish ecology indicate that species segregate on the basis of depth, substrate choice, and food type, it is common for West Coast anglers fishing for rockfish to catch five to ten species in one afternoon, fishing in the same place with just one type of bait.

Some of the rockfish species that are fished commercially are (or were) enormously abundant. In the North Atlantic, redfish supported large fisheries until their populations were depleted in recent years by huge factory trawlers. A similar species, the Pacific Ocean perch, was estimated to have a population in the Gulf of Alaska alone that weighed over 1 billion kilograms when it was discovered by commercial fishermen (mainly Japanese). In just six years (1963–1969), the population was reduced over 60 percent by fishing and it still has not recovered, even though fishing pressure is much less today. In the late 1970s, it was discovered that widow rockfish off the coasts of California, Oregon, and Washington were another abundant rockfish that had not been harvested. They were overlooked previously because they are active mainly at night and can only be caught in numbers after dark. Harvey Gunderson of the University of Washington has shown that this species is also headed for a population crash like that of the Pacific Ocean perch and that the regulatory agencies seem to lack the power to stop this from happening.

The reasons some fishes seem to have such difficulty recovering from overfishing is a bit of a mystery because even a small number of females can produce enormous numbers of eggs. Presumably, a high survival rate of the larvae is possible only under unusual circumstances and population recoveries may have to wait for those circumstances to occur. It is significant, however, that most fishes that live on rocky bottoms have pelagic larvae, many of which get shipped to new areas by oceanic currents. This is the reason that artificial reefs are colonized so quickly. Nevertheless, it may take a few years for fishing over them to become worthwhile because the colonizers have to grow to a catchable size. This may take a surprisingly long time. Many rockfish are seven to ten years old before they enter the fishery, and really large individuals may be seventy or more years old. Off southern California, fishes and invertebrates have colonized the large underwater pipes that carry sewage several kilometers offshore, over muddy bottoms. The fishes that are abundant on these long narrow "reefs" are uncommon elsewhere in the region.

KELP FORESTS

Kelp are giant species of brown algae that form underwater forests in the shallow waters of cold temperate seas. Off the Atlantic coast, these forests are comparatively stunted; individual plants rarely exceed 7 meters in length and a canopy may form only 1 or 2 meters off the bottom. Kelp forests are not found south of Cape Cod. In contrast, off the Pacific coast, kelp plants (of different species) commonly grow to lengths of 30 meters or more, forming beds that truly resemble terrestrial forests, with canopies 20 or more meters above the bottom. Such kelp forests provide a protected environment for animals of all kinds, as they dampen the swells and provide a diversity of habitats. As a consequence, the richness of life in a kelp bed can rival that of a tropical rainforest.

Although kelp forests are an important and widespread habitat, no fishes are adapted for living solely in association with kelp. By and large the fishes are the same ones that occur in nearby rocky reefs. There are a number of reasons for this but one of the most important is simply that kelp needs rocks to anchor the plants, so rocky reef fishes are naturally the ones most available to colonize a new kelp bed. Another factor is that kelp forests are not permanent. Kelp grow incredibly fast, so a "mature" forest may be only two or three years old. Such a forest can disappear quickly if an especially severe storm rips out the plants, if the plants are killed by unusual oceanographic conditions (e.g., high temperatures), or if the plants are chewed up by an infestation of invertebrates, such as sea urchins.

The most studied of kelp forests are those of the giant kelp off the coast of southern California, where it has been shown that a considerable amount of specialization exists among the fishes according to how they use the kelp plants. Distinct groups of fishes are associated with the holdfasts, stalks and fronds, and bottom; within these groups there is subdivision by both space use and type of food. Thus, on the holdfast (the complex plant "root") small eel-like fishes (blennies, pricklebacks) can be found in the openings between roots, whereas clingfishes and snailfishes cling to the plant itself. Roaming just above the substrate, sucking in invertebrates of various sizes, are deep-bodied species such as black surfperch, opaleye, halfmoon, garibaldi, and California sheepshead. Still further out are schools of plankton-feeding fish such as blacksmith and olive rockfish which use the kelp as cover when they are not feeding. Preying on such species are giant seabass and kelp bass. Considerable segregation occurs among the species according to size and kind of prey taken, as well as time of day feeding takes place.

One of the most fascinating aspects of kelp forests is the importance of biotic interactions in their maintenance. For example, sea urchins have been shown on both coasts to be important grazers of young kelp plants; if the urchins are too abundant, kelp has a difficult time getting established and forests do not develop. On the West Coast, sea otters are an important predator on sea urchins so kelp beds are more likely to become established where sea otters are common. This means that fish are more abundant where there are sea otters because there is more habitat for them. Studies of Indian middens in Alaska have shown that fish were important in the diets of the people after a village became established, but that shellfish gradually became more important, presumably because the Indians had killed all the local sea otters. In southern California, sheepshead, a large wrasse (Labridae), may also be important for kelp beds because it feeds preferentially on sea urchins. Because sheepshead are a favorite target for people who spear fish, taking of even a few large individuals from an area may affect the future of local kelp forests.

Another wrasse, the senorita, affects the health of mature kelp forests because it preys selectively on a small amphipod that feeds on kelp fronds. B.B. Bernstein and N. Jung have found that if the senoritas are removed, the amphipods may become too abundant and kill the kelp through their grazing. Curiously, senoritas can harm newly established kelp because they feed on small invertebrates that attach to the fronds; when the kelp is young, the leaves are thin, and the feeding senoritas take bites of frond along with their prey. Excessive feeding of this type can kill the young plants.

SOFT BOTTOMS

Much of the ocean is relatively flat and featureless, especially on the continental shelves, and is covered with layers of silt, sand, broken shells, and other materials. Habitat diversity is low, so usually only two to five species of fish are abundant in any given area, although intensive sampling over a period of years may yield a total list of fifty to sixty species. Fish can be very abundant, however, and support major fisheries, especially for flatfish (flounder, halibut, sole) and various members of the codfish family (cod, hake, pollock, haddock). These can be caught in large numbers because of the ease with which trawls can be dragged across the flat bottoms and because the fish often concentrate in certain areas for feeding or spawning. The catches of trawlers are typically dominated by just one or two species; hauls containing 90 percent or more of one species are not unusual. The reason for this is that the different species show distinct preferences for different bottom types, depths, and oceanographic conditions.

Bottom type has long been an important source of information to fishermen. Before the days of electronic fish finders, fishing boats would often carry a weighted line on board, with a pocket for tallow at the end of the weight. Bottom type—and its suitability for fishing—was determined by examining the material stuck to the tallow, after the line had been lowered to the bottom. Bigelow and Schroeder in their classic 1953 book *Fishes of the Gulf of Maine* note that true cod and hake have such distinct preferences for bottom types that "a long line set from a hard patch out over the soft surrounding ground will often catch cod at one end, hake at the other."

The association of fishes with different bottom types is largely related to feeding habits. Fishes found over silty bottoms are generally specialized for feeding on invertebrates that burrow into the bottom. However, the relationship between fish distribution and bottom type may also be related to other factors, because bottom type may vary with depth, presence of currents, and other factors.

Depth may be an especially important factor affecting distribution. Most fishes have relatively narrow depth ranges, in part because of physiological difficulties of adjusting to pressure changes that accompany depth changes. On soft bottoms, the number of species gradually decreases between 50 and 2,000 meters, as does the total number and biomass of fish, although average size of individuals tends to increase. At depths of less than 50 meters, the number of species and their abundance reflects seasonal and oceanographic conditions. In some areas there is a general decrease shoreward on soft bottoms, presumably because of the unstable conditions that can exist in shallow water and because of the effects of pollutants. Thus, A. V. Tyler found that on the Atlantic coast, the percentage of species that are resident year-round in bays increases towards the north, because temperature and other factors vary less the further north you go. In Pasamoquoddy Bay, New Brunswick, for example, over 30 percent of the fish species are year-round residents. In Chesapeake Bay, in contrast, the fish fauna is made up almost entirely of species that move in and out on a seasonal basis.

The limited range of temperatures at which most fish live is dramatically illustrated by the Atlantic tilefish, a fish that lives in burrows in soft bottoms at depths between 90 and 200 meters. Usually temperatures at these depths do not stray from 8° to 12° C. If unusual conditions cause the temperatures to rise, tilefish may die in enormous numbers and the fishery for them collapses.

Natural events often put severe stress on fish populations, so the added stress of pollutants may be the "last straw" that causes die-offs or emigration. Off the coast of New York, large areas of underwater wastelands have been created by the dumping of garbage, where conditions are so extreme (especially the lack of oxygen) that few fish can live there. Most bays associated with large cities suffer from depleted fish faunas; the number of species and individuals are less and the incidence of disease is much higher than under natural conditions. In Boston Harbor, not only are there fewer species than might be expected, but the incidence of carcinomas is very high in at least one common species, winter flounder. Off the coast of Los Angeles, the diversity of fishes is reduced, especially near sewage discharge sites, and the fauna has shifted from one associated with sand bottoms to one associated with silt bottoms, because of the accumulation of organic matter. Some of these soft-bottom species, such as white croaker and dover sole, are actually attracted to the outfalls, but they are afflicted with tumors, lesions, and fin erosion as a consequence.

SEA GRASS FLATS

On shallow muddy bottoms, where the water is warm and clear, meadows of sea grass may develop. They often grade into salt marshes or mangrove swamps in intertidal areas. There are thirty-five to fifty species of grass that can make up sea grass beds, but in temperate areas the species is usually eelgrass. Sea grass beds are among the most productive plant communities known, rivaling cultivated corn and plankton blooms for the amount of plant matter produced through time. This plant matter encourages invertebrate grazers and detritus feeders, which in turn are consumed by fish (as well as other creatures). Sea grass beds consequently support large populations of fish. However, because sea grass flats are in such shallow water, they provide cover (especially from predatory birds) only for small fishes.

The dominant fishes of sea grass beds are thus juvenile fishes or adults that are less than 20 centimeters long. The diversity of fishes is further reduced by seasonal and daily fluctuations in temperature and salinity that can characterize the beds because of their location in such shallow, clear water. Not surprisingly, the fishes tend to be the same species found in nearby estuaries and bays. S. M. Adams found this to be true of North Carolina eelgrass beds he studied; he collected only twenty-four species more than once in a year of intense sampling. A majority of the fish he collected were juvenile pinfish, a common coastal species. The pattern was similar for a New York eelgrass bed, only the dominant fishes were Atlantic silverside and fourspine stickleback.

The number of fish in a sea grass bed varies considerably with time of day and season. The key factor for fishes seems to be temperature, which can be quite high on warm summer days. In the beds studied by Adams, fish numbers in summer were highest at night, when temperatures were lowest. Numbers of fish were nevertheless highest in summer, presumably because of the high populations of food organisms and the high density of the grass itself which provided cover. This also reflects the importance of sea grass flats as nursery grounds for young of large marine fishes. When the shallow water cools in winter, many eelgrass fishes may move into the warmer water present at greater depths.

OPEN WATER

Although a large majority of species that live in association with the continental shelf live on or close to the bottom, probably the majority of individuals belong to those species that inhabit the open water, especially herrings, sardines, menhaden, anchovies, and their predators. These fishes typically school in enormous numbers within a few kilometers of the coast and are easy to exploit not only by fishermen but by marine mammals and birds. The fortunes of entire coastal communities may rise and fall according to the abundance of just one or two pelagic species; the most famous example is probably Monterey, California, where the canneries of John Steinbeck's famous *Cannery Row* all went out of business when the sardine populations collapsed. This example also demonstrated dramatically that it is even possible to overfish a seemingly limitless population of fishes. This lesson has not been well learned, however, and many coastal pelagic fishes continue to be overfished.

As indicated, the most abundant inshore pelagic fishes are members of the families Clupeidae (herrings, menhaden, sardines, etc.) and Engraulidae (anchovies). These are all small silvery fishes which feed on the plankton that thrives in the nutrient-rich coastal waters. Some species, such as menhaden and anchovies, use their fine, closely spaced gill rakers to filter small plant and animal plankton from the water. Herrings and sardines may also filter feed, but tend to pick out individual zooplankton from the water. Curiously, the filter-feeding anchovies are among the smallest pelagic fishes, yet the largest pelagic fishes, whale sharks and basking sharks, are also filter feeders.

Naturally, the pelagic plankton feeders attract many predators: sharks, tuna, mackerel, jacks, billfish, and salmon. These fishes have developed many styles of feeding, often aimed at disrupting the schools so individual fish can be picked out. Tuna, mackerel, salmon, and jacks use a rapid approach to surprise the schools and confuse the small fishes further because they typically attack in schools. Often they will attack a school from below, driving it to the surface, where the fish leap out of the water in an effort to escape. This serves to attract seabirds that plunge down from above, further confusing the schooling planktivores. Such attacks probably cease mainly when the predators are satiated. Swordfish and marlin have a slashing mode of attack, swimming through a school and stunning as many fish as possible with their long bills. They then go back to feed at leisure on the injured fish. Marlin will even attack small tuna; it is common to find tuna in their stomachs with spear wounds in their sides. Some sharks seem to use a herding approach, swirling isolated pockets of prey into a confused ball of fish that can be eaten easily. Thresher sharks, with their long tails to use as sweeps, seem to be specifically adapted for feeding in this manner.

It has long been known that pelagic fishes make regular seasonal movements and may migrate long distances. Yet it is often not obvious just what the fish are responding to when they make these movements, because, to us, their environment seems uniform in its characteristics.

In fact, the fishes orient very strongly to gradients of temperature, light, and other oceanographic features.

Temperature is particularly important. For example, in the northern hemisphere, salmon tend to be the dominant pelagic predator in water colder than 8°–10° C, whereas tuna and billfishes are dominant in warmer water. Thus as oceanic water warms up in the summer, tunas and billfishes move north and retreat south as it cools. Fishing times for them can thus be predicted to a certain extent by keeping track of water temperatures. Off the California coast, the water is generally quite cold, thanks to the California current that flows down from the Gulf of Alaska. The strength of this current varies from year to year and in weak years, dramatic changes in the fish fauna can be observed, even though temperatures may only increase 2° or 3° C. Fisheries for yellowtail and barracuda may suddenly develop, whereas salmon fisheries may decline. Ichthyologists start accumulating new records of "tropical" reef fishes off the California coast. Reproduction of abundant species may be affected; the fishery for sardines off California collapsed in part because the overfishing occurred at a time when temperatures were too low for adequate reproduction.

Perhaps because the pelagic environment is so devoid of visual stimuli, most fishes there are not only social (usually schooling) but many associate with other species on a regular basis. Schools containing two or more species of fish are not unusual; this may be one way uncommon species can persist, by schooling with common ones. Pelagic sharks are often accompanied by pilot fish, which apparently get a free ride in the shark's "slipstream." Sharksuckers or remoras carry this a step further and actually attach to the sharks with a sucker on the top of their heads. Less well understood is the relationship between porpoises and yellowfin tuna. These two species are commonly found together, so much so that the fishery for the tuna depends on fishermen setting their nets around schools of porpoise in order to catch the tuna. It is assumed that the tuna find fish to eat more efficiently by following porpoise around, because they can locate prey through the use of their echolocation abilities. At the present time, this association is bad for both species, because fishermen can find the tuna more easily by tracking porpoises and many of the porpoises die when caught in the nets. Tuna boats of U.S. registry are required to use special nets that allow the porpoises to escape, but most tuna boats of other countries do not use the improved nets because they are more difficult to handle than the standard nets.

sand shark

sucking disk on
dorsal surface of
remora head

remora

Figure 13-2.   The remora gets a free ride from the sharks to which it attaches
itself; in return the remora cleans parasites from its host.

PROJECTS

1. If you are an experienced scuba diver, organize a group of divers
to swim transects at your favorite diving spot, to monitor the fish populations. Transects should be set up so they are easy to locate and easy
to complete on a normal dive. The transects should be made several
times a year. Cards that can be used underwater to identify common
fishes are becoming increasingly available; check your local dive shop
for availability. Underwater slates for taking notes are also available.

2. If you scuba dive but do not wish to swim transects, keep careful
notes of what you observe on each dive. Learn to identify the common
fishes and note in a general way how abundant they are each time. Keep
track of water temperature and clarity as well, to see how these affect
your observations.

3. Go fishing over a rocky reef and fish with a variety of hook sizes
and baits. Be sure to fish much of the time close to or on the bottom.
Keep track of the fishes you catch and how you caught them. Examine

stomach contents and record your findings in a notebook. If you fish in the same place over a number of years, such records yield much interesting and useful information, especially if you keep track of such factors as temperature and water clarity as well.

4. Go to a local fish market on the coast and identify the common fishes. If you do this on a regular basis, keep track of the changes in available fishes.

CHAPTER XIV

Tropical Reefs

For years I lectured on the fishes of tropical reefs in my classes, using colorful slides provided by colleagues, before I actually visited one. Even so, I was hardly prepared for the beauty and activity found on a tropical reef when I finally got a chance to snorkel over one. Photographs, no matter how good, simply could not prepare me for the sheer numbers of fish, seemingly in constant motion, with iridescent colors that changed constantly in the streaks of wave-dappled sunlight. After my initial astonishment wore off, I finally was able to focus on a coral knob, where a tiny blue wrasse was energetically dancing about and picking parasites from a line of lethargic looking yellow snappers. It was wonderful to discover how beautiful and intricate this "cleaning" activity was, after lecturing on it so often in ponderous detail.

My experiences on that Sri Lankan reef are very modern ones, made possible by the development of face masks, snorkels, and scuba gear. Invention of this gear has been the best thing to happen to aquatic naturalists since the invention of the fishhook. Studies of tropical reef organisms have burgeoned in recent years as biologists have become able to work directly with the incredible array of organisms that live in the spectacularly beautiful settings reefs provide. The shimmering clouds of multicolored fishes alone are enough to excite the wonder of anyone who snorkels or dives over a reef. The diversity of oddly shaped and brightly colored fishes living on reefs truly is astonishing; nearly 40 percent of the 21,000-plus fish species are associated with tropical reefs in one way or another.

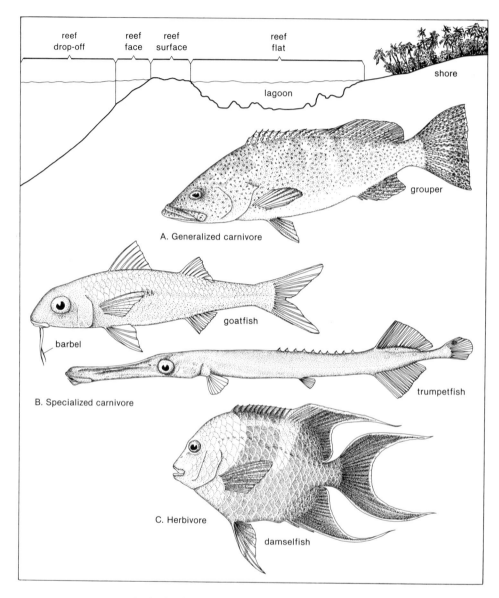

reef
drop-off

reef
face

reef
surface

reef
flat

shore

lagoon

grouper

A. Generalized carnivore

goatfish

barbel

B. Specialized carnivore

trumpetfish

C. Herbivore

damselfish

Figure 14-1. The body shapes of tropical reef fishes reflect the wide variety of specialized methods they use for obtaining food.

A typical reef may house between 250 and 2,200 species. Every crack and cranny seems to contain a fish; C. L. Smith and J. C. Tyler, for example, collected 75 *species* and numerous individuals from an isolated chunk of coral in the Caribbean that was only 3 meters in diameter. As might be expected in such a crowded environment, many of the fishes are highly specialized in their feeding habits, choice of habitat, and breeding behavior. Studies of these specializations are exciting but surprisingly few; they are a good place for a patient amateur to make significant contributions to science. Developing an appreciation for the complexity and beauty of tropical reefs is also important because the same factors that have opened them up to study, the development of diving technology and modern rapid transportation systems, have also made them vulnerable to exploitation. Excessive removal of fish, coral, and invertebrates is changing, even destroying, many reef communities. Unfortunately, we have a poor understanding of how much exploitation most reefs can take or how to conserve the reef biota for future use and enjoyment.

This chapter can provide only a brief and tantalizing glimpse of reef fishes by describing their environment, the types of fishes present, interactions among species, daily activity patterns of the fishes, and fish communities.

THE REEF ENVIRONMENT

Tropical reefs are found in warm, clear water between the latitudes of 30 degrees north and 30 degrees south. For reefs constructed by corals, water temperatures are usually well above 18° C and underwater visibility ranges from 10 to 20 meters. Although coral reefs are the most famous and extensive reefs, many other reefs are constructed by algae that produce hard calcium carbonate. Along continental margins, especially in cooler areas, the reef fauna may simply inhabit rocky areas. In the Gulf of Mexico, a limited assemblage of tropical reef organisms even develops around many offshore oil drilling platforms.

Tropical reefs occur in four widely separated oceanic regions: the Indo-Pacific, the Eastern Pacific, the West Indian, and the West African. Each of these areas has its own distinctive fish fauna and consequently requires different field guides. The *Indo-Pacific* is by far the largest of the regions because it includes most of the tropical Pacific and Indian oceans and associated coastlines of Asia and Africa. Despite its vast size, the fish faunas from one end to the next are surprisingly similar. Reefs in the more isolated areas, such as the Hawaiian Islands, differ from those in the center (around Indonesia and Australia) mostly by containing fewer species. They may contain a few endemic species as well. The *Eastern Pacific* region encompasses the scattered reefs found along the Pacific coast of Central and South America, including the rocky reefs of the Gulf of California. The *West Indian* region is the most accessible to inhabitants of eastern North America and includes the islands of the Caribbean, the Bahamas, the tip of Florida, and Bermuda. It is second only to the Indo-Pacific region in its richness of fishes. The *West African* region is the smallest, least rich in species, and poorest known of the reef regions, because it consists of small reefs scattered along the west coast of Africa.

Despite the diversity and wide extent of reef faunas, tropical reefs have a common structural pattern (if not viewed too strictly) that is generally expressed as six habitat zones. Each zone typically contains fishes adapted for its particular conditions, as well as more wide ranging species. The six zones most often recognized (but not present on every reef) are the off-reef floor, reef drop-off, reef face, reef surface, reef flat, and lagoon.

Off-reef floor.   Many reefs are interspersed with flat, sandy areas. In the immediate vicinity of the reef they are barren looking but out a few meters they support beds of sea grass. The moat of sand around each reef is created by grazing fish and invertebrates that are unwilling to venture too far from the solid cover a reef provides because they can be snapped up by fast-moving predatory fishes that circle every reef. The sand itself contains hazards for grazers, such as sand divers (Synodontidae), predatory fish that bury themselves in the sand, waiting for unsuspecting prey.

Reef drop-off.   More often than not, reefs do not end in sandy flats but in steep drop-offs into the depths. Living corals and algae gradually become less abundant as depth increases and the diversity of fishes decreases. However, the upper 50 to 60 meters may be favored by large numbers of fishes that use the available cover but still remain close to the plankton they feed upon. There is frequently good fishing along the reef drop-off because of the predatory fishes that patrol it, looking for prey.

Reef face.   This is the place where snorkeling is at its best. The water is shallow and well lighted, the surge from the waves moderate, the bottom carpeted with a rich panoply of corals and invertebrates, among which swim hundreds of fish of all shapes, sizes, and colors. The visible fishes are only a portion of the total fish fauna; there are many small species hiding in the cracks and crevices, whereas nocturnal species lurk in the caves and hollows. Usually the first fishes seen by a novice are large schools of small fishes such as damselfishes (Pomacentridae) or tangs (surgeonfishes, Acanthuridae), which hover and maneuver above coral heads. Next the numerous brightly patterned butterfly fishes (Chaetodontidae) and angelfishes (Pomacanthidae) are seen, followed by such large fishes as parrotfishes (Scaridae) and groupers (Serranidae). The more you look, the more fishes you see, including the many carefully hidden flounders, eels, and scorpionfishes (Scorpaenidae).

Reef surface.   This area may be good for snorkeling in a few of the larger pools, but generally it is too shallow and has too much surge from the waves to be comfortable, or safe, for snorkeling. It is, however, rich in life and many small fishes can be seen as you walk carefully across the shallower areas. Caution is advised not only to avoid damaging the invertebrates but to avoid stepping on the spines of such highly poisonous fishes as the stonefish.

Reef flat.   Behind the main reef is often a protected area, frequently including a lagoon, which has large expanses of flat sandy bottom containing isolated chunks of coral. These areas are very easy and safe to snorkel over and the number of fishes on the isolated coral heads is sometimes quite high.

Lagoon.   The lagoon of an atoll can also be a delightful place to snorkel or dive, especially where there are small reefs in the middle of it. Unless there is a good surge feeding the lagoon, the diversity of fishes may be somewhat less than on the more exposed reef surface, although the area bordering the lagoon near a channel entrance may have the highest diversity of any reef zone.

Because diving on reefs is done on calm, pleasant days and because the same (or seemingly the same) fishes can be seen in the same areas for long periods of time, reefs are often thought of as models of stability. In fact they are changing constantly. Corals and calcareous algae are growing continually, filling in old spaces and creating new ones. Coral heads fall off or become covered with sand. Tropical hurricanes, on a highly irregular basis, may knock big holes in a reef and roll the chunks into lagoons. Thus on a local scale, the habitat is changeable, although on the much broader scale it is fairly constant (as long as ocean temperatures do not change due to global warming).

REEF FISHES

E. S. Hobson of the National Marine Fisheries Service has devised a rather convenient way of classifying reef fishes according to their method of feeding, which will be used here. There are three broad categories in his system: herbivores, generalized carnivores, and specialized carnivores.

Herbivores.   Algae-eating fishes make up less than a quarter of the species on tropical reefs, but they are among the most important in terms of numbers and effects on the reef environment. The most noticeable of the herbivores are usually damselfishes and parrotfishes. Herbivorous damselfishes are frequently less brightly colored than more carnivorous members of the family. Most of the herbivorous damselfishes are gardeners, carefully maintaining a "crop" of algae on a small section of reef which they defend vigorously from all other fishes. Because algal growth is thicker inside damselfish territories than outside them, large roving grazers, such as parrotfishes, will overwhelm a defender by moving in as a school in order to consume the territory's contents. The demand for algae is so great on a reef that herbivorous fish and invertebrates keep the algae cropped down to a thin mat a few millimeters thick and maintain a sea grass-free ring in sandy areas around reefs as much as 10 meters wide. When herbivores are kept away from a section of reef, it quickly grows a luxuriant coat of algae, made up of many species. The grazing of the herbivores, however, accounts for much of the high productivity of the reefs, because algae must grow rapidly to keep up with the grazing. Even though it is barely visible to humans, the thin algal coat on reefs converts the energy of sunlight into the thousands of grazing fish we do observe. The grazers then become food for the carnivores.

Generalized carnivores.   These are rover-predators in body shape, with large mouths designed for capturing other fishes and mobile invertebrates. They generally cruise about the reef, taking their prey by surprise. Different body characteristics and different modes of prey capture are employed by species active at different times of day. The basslike groupers have body colors and behavior that allow them to blend in with the reef. Many of them take their prey from ambush during the day, waiting for a fish to make a defensive mistake, such as foraging too far from a school or from cover. Crepuscular predators become active as light levels in the water dim in the evening or increase in the morning because the distance from which their prey can spot them is greatly reduced. Some snappers (Lutjanidae), move slowly above the reef, sneaking up on their prey and then seizing them with a sudden rush. Others, such as the streamlined jacks (Carangidae), cruise along the reef fringe, darting in and out at high speeds in order to surprise schooling fishes. Nocturnal predators have large eyes and feed mainly on night-active invertebrates. Many, such as squirrelfishes (Holocentridae), are bright red in color, which offers concealment (see chapter 2).

Specialized carnivores.   These are the most visible fishes on the reefs. Their specializations are so many and so bizarre they can only be hinted at here. Hobson divides specialized carnivores into seven broad types: ambushers, water column stalkers, crevice predators, concealed prey feeders, diurnal predators on benthic invertebrates, cleaners, and diurnal planktivores. **Ambushers** are so good at concealing themselves they are frequently overlooked by both prey and human observers. Good examples are lizardfishes (Synodontidae), scorpionfishes, and peacock flounders (Bothidae). Many of these fishes seem to have extraordinarily bright color patterns when removed from the water, yet these same patterns and colors blend in well with the brilliant, kaleidoscopic reef background. Like the ambushers, **water column stalkers** try to capture prey by making themselves as invisible to the prey as possible. They do so by being silvery and elongate, with long snouts full of sharp teeth. They drift in the water column and from a head-on perspective appear much smaller than they actually are. This allows them to approach small fish close enough to grab them with a sudden lunge. The most extreme examples are the cornet and trumpetfishes (Fistulariidae) which not only have attenuated snouts but also a long spike behind the tail to reduce shadows. These fishes also will hide behind large parrotfishes foraging on the reef, using them as mobile cover.

Eels are the best examples of **crevice feeders**. They have elongate bodies and narrow heads for penetrating deep into caves and crevices, in pursuit of hiding fishes and invertebrates. **Concealed prey predators** also search crevices, but for smaller prey, using "feelers" of one sort or another. Examples are the abundant goatfishes, which have long chin barbels they can use to probe the reef, as well as flexible mouths they can use to suck up a prey item once detected.

The most conspicuous fishes on reefs are generally **diurnal predators on benthic invertebrates**, such as butterfly fishes, wrasses, parrotfishes, and pufferfishes. They all use vision to find their prey and feed mainly on conspicuous reef invertebrates such as sponges, corals, tunicates, sea urchins, and molluscs. As might be expected, the invertebrates do their best to avoid being eaten by the fishes and have evolved a formidable array of defenses including spines, toxins, bad taste, and heavy armor. In a classic case of a natural "arms race" the fishes have had to become increasingly specialized to overcome invertebrate defenses. Armor is crushed with strong jaws and beaks (as in puffers and boxfishes) or strong pharyngeal teeth, as in the wrasses. Because most armored invertebrates have to expose soft parts to feed, many fishes, such as some butterflyfishes, sneak up on them and clip off pieces with their small, sharp teeth. Other butterflyfishes and parrotfishes scrape mucus secreted by coral and the associated algae and invertebrates.

A specialized case of benthic feeding is found in the **cleaners**, small fishes that pick parasites and bits of loose and diseased tissue from the sides, fins, and mouths of other fishes. In the Indo-Pacific region this role is played mainly by wrasses whereas in the West Indian region it is played mainly by gobies. It is truly remarkable that similar, highly specialized behavior evolved in fishes from completely different fish families. Many of the cleaners have set stations on the reef over which they flash their bright colors to attract customers. Large and small fish soon are lined up at the station, each going into a trance-like state as it is cleaned. Although this interaction seems to be mutually beneficial, with the customer fishes having parasites removed and the cleaner fish getting meals, it is not so simple. Cleaner fishes apparently also take healthy scales, mucus, and bits of fin and the main attraction for the customer fishes seems to be the tactile stimulation provided by the cleaners. Cleaner fishes pick off mainly the larger parasites, so fishes that are regularly "cleaned" tend to have fewer large parasites, but more smaller ones. However, there is evidence that wounds treated by cleaner fishes are more likely to heal than untreated wounds.

The preceding groups of specialized carnivores feed mainly on the reef itself but another group, the **diurnal planktivores**, uses the reef mainly for cover. These fish forage off the reef for small zooplankton. This results in a curious conflict in their body design because they need to be deep bodied, with large eyes and small snouts, to capture the plankton and yet streamlined to be able to speed back to the reef if a predator comes along. As a result, they are moderately deep bodied, with narrow caudal peduncles and deeply forked tails. These features have evolved in plankton feeding fishes from a number of different families such as damselfishes (Pomacentridae), surgeonfishes (Acanthuridae), and small seabasses (Serranidae). Typically, these fishes hover just over the reef in small clouds, darting back and forth as they feed, and diving into cover at every large passing shadow. The feeding habits of these planktivores has the effect of increasing the productivity of the reef because the planktivores deposit their feces on the reef, providing nutrients for algae and other organisms.

SPECIES INTERACTIONS

In a system as packed with individuals and species as a tropical reef, interactions among fishes are bound to be numerous and complex. Examples of competition and predation have already been mentioned, so this section will concentrate on two types of interactions particularly characteristic of complex systems: symbiosis and mimicry.

Symbiosis.   The types of symbiosis most visible on reefs are mutualism and commensalism. The classic case of **mutualism** (where both species benefit) is often considered to be the cleaner fish-customer fish interaction, although, as pointed out, recent research has shown the interaction is not as straightforward as once thought. A better example seems to be the interaction between anemone fish (Pomacentridae) and anemones, attached invertebrates that use poison stings to capture prey. The anemone fish acquire immunity to the stings and then live in and under anemones, using them to protect themselves and their nests. In return, the anemone fish captures invertebrates and small fishes and shares its meals with its anemone, keeping it healthy and growing. Another example is that of the gobies (Gobiidae) that share burrows built by shrimp. The shrimp apparently encourage the gobies because the good eyesight of the fish provides the shrimp with an early warning system for the approach of predators.

Examples of **commensalism** (where one species benefits from the actions of another, but the benefactor neither benefits nor is harmed by the other species) are also common, such as the small fishes that hide inside the hollow tubes of sponges or among the spines of sea urchins. Among fishes, following behavior is a common type of commensalism, where small predatory fishes closely follow large herbivores (e.g., parrotfish) in order to snap up the invertebrates disturbed by the grazing. Foraging moray eels may be followed by other fish-eaters as well, hoping to grab small fishes that emerge from their crevices while escaping from the eels.

Mimicry.   The more coral reef fishes are studied, the more examples of mimicry are uncovered. Best known is the case of the sabre-toothed blenny (Blenniidae) which mimics the cleaner wrasse. The blenny lures large fish to it by behaving like a cleaner, but then takes bites out of the fins rather than picking off parasites. Usually only naive or young fish fall for its tactics, however. A remarkable case recently discovered by John McCosker, director of San Francisco's Steinhart Aquarium, is that of the marine betta, which mimics a spotted moray eel to escape predators. When pursued by a predator, it dives into cover but leaves its rear half exposed. The dorsal, anal, and tail fins are merged together and a large eye spot on the dorsal is exposed; to the pursuer, the betta has suddenly been transformed into the head of a large aggressive moray eel!

ACTIVITY PATTERNS

Biologists who have spent much time watching reef fishes are impressed with the dramatic differences between daytime and nighttime fish activity on reefs. Essentially, there are nocturnal and diurnal faunas, a division that contributes to the diversity of reef fishes. During the day, the reef is busy with the activity of the familiar fishes, as the brightly colored carnivores pursue their prey and the herbivores graze and defend territories. Actively hovering and feeding above the reef are the small planktivores, closely watched by patrolling piscivores. Much less active and much closer to the reef are schools of nocturnal planktivores (many of which are damselfishes as well), waiting for night to come so they can leave the reef and feed on the larger planktonic organisms that are avail-

Figure 14-2. The spotted moray eel (*top*), a predator on small fishes, spends much of its time in reef crevices, with only its head exposed. The marine betta fools its predators by mimicking the head of the spotted moray using its tail, spotted pattern, eyespot, and behavior.

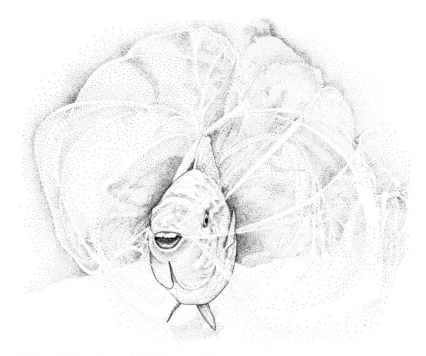

Figure 14-3. Parrotfish hide in reef crevices at night and may make themselves more difficult for predatory moray eels to find by secreting a loose envelope of mucus about themselves.

able at night. Deep within the crevices of the reef are other nocturnal foragers, large-eyed and red in color, such as squirrelfishes (Holocentridae), cardinalfishes (Apogonidae), and bigeyes (Priacanthidae).

As the light begins to dim in the evening, the diurnal fishes begin to seek cover, often swimming some distance to a favorite resting place. Once under cover, small parrotfishes may secrete a cocoon of mucus about themselves, apparently to fool night-foraging morays. The diurnal planktivores drop down close to the reef, schooling closely together. According to E. Hobson, on Hawaiian reefs most of these fishes are in place fifteen to twenty minutes after sunset. Next, a period of twenty minutes or so of low activity begins. Presumably most fishes avoid being out during this period of dim light because they are so vulnerable to predators, which are actively seeking prey at this time. Once darkness settles in, the nocturnal planktivores boil from their hiding places and quickly swim up into the water column. On the surface of the reef, only a few fishes are moving about in the dark, mainly eels and goatfishes seeking invertebrates crawling over the reef. At dawn, the evening events reverse themselves and the diurnal fishes quickly become active.

COMMUNITY STRUCTURE

The first impression of anyone snorkeling over a tropical reef is that, compared to most other aquatic environments, it is crowded with fishes. Closer examination reveals that it is even more crowded than expected, as small and cryptic species are discovered and as fishes are observed in caves and crevices. Trying to explain the extraordinary diversity and numbers of tropical reef fishes is an important and exciting area of research among ecologists. Some of the explanations proposed will be discussed briefly here, dangerously oversimplified.

The classical explanation for the high diversity of reef fishes is that they have evolved together over a long period of time and have developed highly specialized means of dividing up the resources (algae, invertebrates, cover, etc.) of the reef. In this view the reef is a very ordered place, with competition rarely occurring; each species has its own specialized niche that has minimal overlap with the niches of other species. Indeed, the many extraordinary specializations of reef fishes in feeding and other uses of the reef seem to support this view. The problem is that there are many similar species on reefs that do not seem to be sharply segregated and do seem to need the same resources for their existence, especially territory for feeding and hiding.

One solution to this dilemma has been proposed by Peter Sale. Dr. Sale noted that most reef fishes have extremely long spawning seasons— some spawn all year round on a nearly daily basis—and that most produce planktonic eggs and larvae. The latter fact accounts in good part for the wide distribution of species on scattered reefs, as they are carried there by ocean currents. Planktonic larvae have notoriously low survival rates because they are eaten by invertebrates and fishes, swept away into environments where they starve to death, or are devoured by resident fish when they finally settle out on a reef. Sale proposed that for a fish to survive, it has to settle out on a space not occupied by another fish and that the probability of this is so low that in effect a lottery exists among the larval fishes. The winner gets the space, regardless of species. One problem with this idea is that the larvae of each species must have an equal probability of settling in a given spot, which seems unlikely.

By whatever the mechanism, coexistence of species may be permitted more easily if the environment is not saturated with fish, despite its appearance of being so. This way competition may not be a problem because there are plenty of resources to go around. Some experiments have shown that fish can be added to a section of reef and not seem to increase the mortality rates of fishes already present. One way fish populations are kept at low levels is through constant harvest by predators, although this has not been demonstrated for reef fishes. Periodic disasters, such as tropical storms, may also reduce fish populations, although this happens too infrequently to be a major cause of the patterns observed. Robert K. Warner of the University of California, Santa Barbara, has suggested that the mortality rates of larval fishes are so high and irregular that rarely are enough fish recruited to a reef to keep the populations at a saturation level. He argues that reef fishes live a long time so that species are maintained by the few times during which conditions are right for their larvae to survive in abundance. This argument is backed up by studies showing that most individuals of some species took up residence on the reef at about the same time. Presumably, the diversity of fishes on reefs is maintained by a variety of mechanisms working together but the precise elucidation of these mechanisms will keep biologists and naturalists employed indefinitely.

PROJECTS

A casual visitor to a reef in Hawaii, the Caribbean, or the Gulf of California is likely to find just the identification of fishes to be a major project. To get a good feel for the life of reef fishes, spend some time watching one small part of a reef, especially if it contains a cleaner station. If you visit reefs on a regular basis, establish some regular snorkeling routes and count the numbers of one or two of the more spectacular forms, such as butterflyfish. You may even find yourself recognizing individual fish, which may stay in the same place for several years in a row. Serious fish watchers will find good ideas in R. and J. Wilson's *Watching Fishes: Life and Behavior on Coral Reefs* (Harper and Row, 1985).

CHAPTER XV

Conservation

The last word in ignorance is the man who says of an animal
or plant: "What good is it?" If the land mechanism as a
whole is good, then every part is good, whether we understand it or not. If the biota, in the course of aeons, has built
something we like but do not understand, then who but a
fool would discard seemingly useless parts? To keep every
cog and wheel is the first precaution of intelligent tinkering.
                            —*Aldo Leopold*, Round River*

In 1989, two other scientists and I completed a report to the California
Department of Fish and Game called *Fish Species of Special Concern
of California*. We evaluated the status of the 113 native freshwater fishes
in the state. We found that:

1. Seven were extinct.

2. Fourteen were officially listed as threatened or endangered.

3. Seven needed immediate listing as threatened or endangered.

4. Nineteen were in serious trouble and would merit listing soon if
their populations continued to decline.

5. Twenty-five had declining populations or naturally limited
ranges but did not appear to be in serious trouble, although their populations needed monitoring.

6. Forty-one appeared to have secure populations.

Although the ecology and conservation of California's native fishes
has been one of my main areas of research for over twenty years, I was
shocked by results that showed only 36 percent of California's unique
fish fauna could be regarded as reasonably safe from human-caused extinction. Unfortunately, our figures were actually conservative. By 1992,

*Published in 1953 (New York: Oxford University Press), p. 147.

new information indicated that one of our "serious trouble" fishes had gone extinct, and that twenty additional fishes probably merited being listed as threatened or endangered.

Although California is showing real leadership in the race towards extinction of native fishes, other states and countries are catching up. Nationwide, a study by the American Fisheries Society showed that more than one-third (343) of 950 species of freshwater fish found in North America were listed as endangered, threatened, or of special concern by at least one state. The U.S. Fish and Wildlife Service listed, in 1990, 93 fishes as being threatened or endangered throughout their range. A view of the picture worldwide is presented in table 1. Overall, a reasonable estimate for 1990 is that 20 percent of the world's freshwater fish species (about 1,800 species) are extinct or needing special protection to keep them from becoming extinct. This number is growing rapidly.

Freshwater fish are reflecting a worldwide trend in the loss of species of all kinds. It is a trend that has only begun to touch marine fishes but they too can expect to be caught up in the rapidly growing "extinction vortex" unless major conservation efforts are made soon. To make these efforts, we need to understand why we should protect obscure fishes, why so many fishes are in such trouble, and what efforts are being made already.

WHY SHOULD WE CARE ABOUT LITTLE FISHES?

The telephone message I found in my mailbox the other day, clearly written in the neat cursive handwriting of the departmental secretary, said simply "The High Rock Spring tui chub is extinct." The message was from a biologist in the California Department of Fish and Game who knew I was one of the handful of people who would be as upset about it as she was. Upon returning the call, I found out the small fish had quietly swum out of existence some time in the past couple of years, following introduction of other fishes into the isolated desert spring in which it lived. But why *should* I, or anyone else, be upset about the loss of a fish I had seen only in photographs? It was, after all, a very ordinary looking minnow that lived in the main water source of a private ranch. It was replaced by two highly edible species. My reasons for concern are at once ethical, aesthetic, and practical.

Ethically, I think it is simply not right to eliminate other species of organisms from this planet, with the possible exception of a few human disease organisms. Other species have as much right to continue to exist

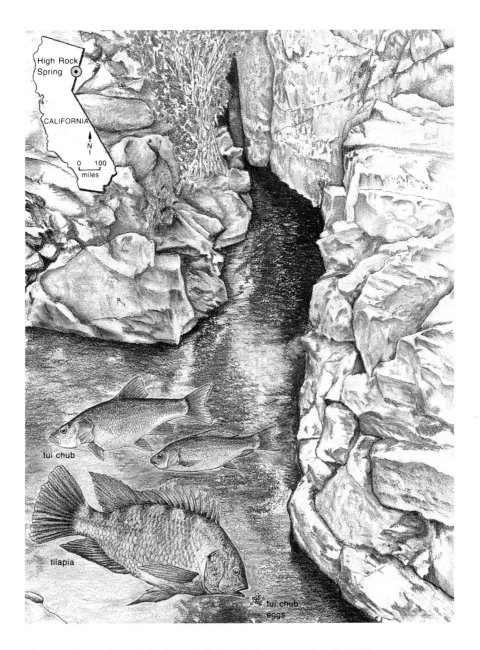

Plate 1. The High Rock Spring tui chub quietly went extinct in 1989.
It lived in a small desert spring in northeastern California and was
apparently eliminated when African tilapia species were introduced into
the spring during a failed fish farming operation. The tilapia presumably
preyed on the eggs and young of the chubs.

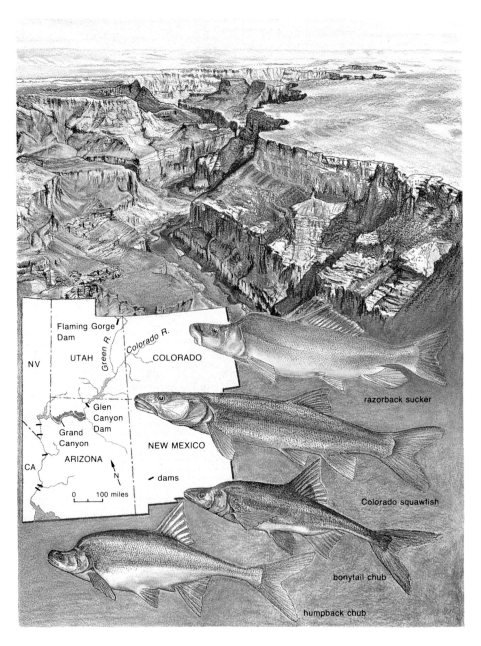

Plate 2. Endangered fishes of the Grand Canyon. Numerous dams on the Colorado River (represented by bars on the map) have so altered the riverine habitats that these highly unusual fishes persist mainly in the Green River, upstream from the canyon itself.

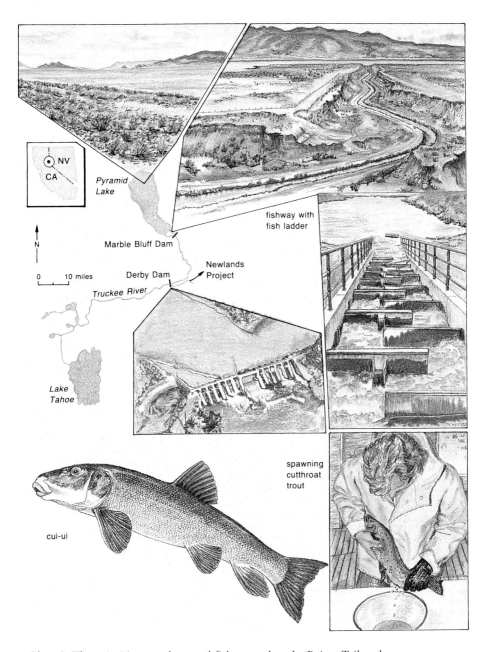

Labels within image:
NV
CA

Pyramid
Lake

fishway with
fish ladder

N

0   10 miles

Marble Bluff Dam

Derby Dam

Newlands
Project

Truckee River

Lake
Tahoe

spawning
cutthroat
trout

cui-ui

Plate 3. The cui-ui is an endangered fish, sacred to the Paiute Tribe, that
lives only in Pyramid Lake, Nevada. It became endangered when Derby
Dam reduced the flow of the Truckee River in which the fish spawn. The
enormous efforts made to preserve this fish, and a native cutthroat trout
as well, include construction of 4.8 kilometers of fishway to provide
access to the river and spawning the fish artificially and raising them in
hatcheries.

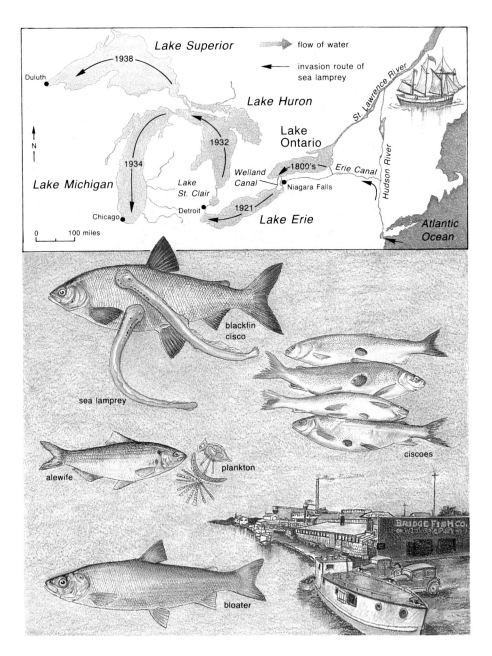

Plate 4. The Great Lakes once contained a diverse assemblage of ciscoes (whitefish), which supported large fisheries. Today, four of the species are already extinct, and three others are endangered. The principal causes were two fish species that invaded through the Welland Canal around Niagara Falls: the alewife (a cisco competitor), and the sea lamprey (a cisco predator). The map shows the rate at which the lamprey managed to invade the lakes and decimate the fisheries.

Plate 5. The Owens pupfish is a small, feisty fish that was rescued from extinction, literally at the last minute. It now thrives only in special sanctuaries created for it and the three other native fishes shown (not drawn to scale). The scene in the lower right shows a historic moment when the rescued pupfish were introduced into the newly created Owens Native Fish Sanctuary.

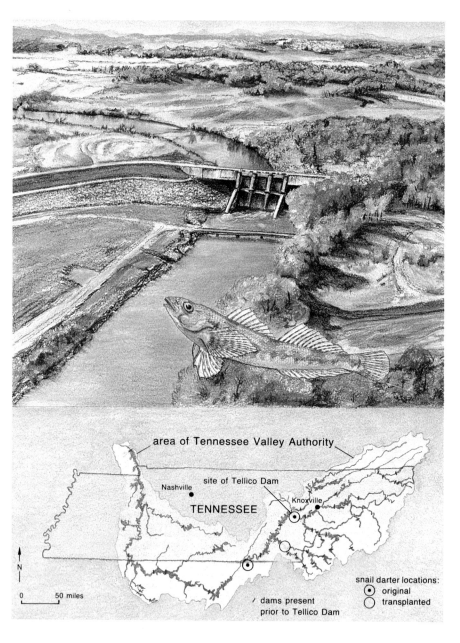

area of Tennessee Valley Authority

site of Tellico Dam

Nashville

TENNESSEE

Knoxville

N

0   50 miles

snail darter locations:
⊙ original
○ transplanted

✓ dams present
prior to Tellico Dam

Plate 6. The snail darter is a small fish of the Little Tennessee River that provided the first major test of the Engangered Species Act, as the darter's existence was thought to be threatened by Tellico Dam, shown under construction. The act was upheld, but the dam was built anyway, as the result of political subterfuge. Fortunately, the darter managed to survive in transplanted populations and in previously unknown populations.

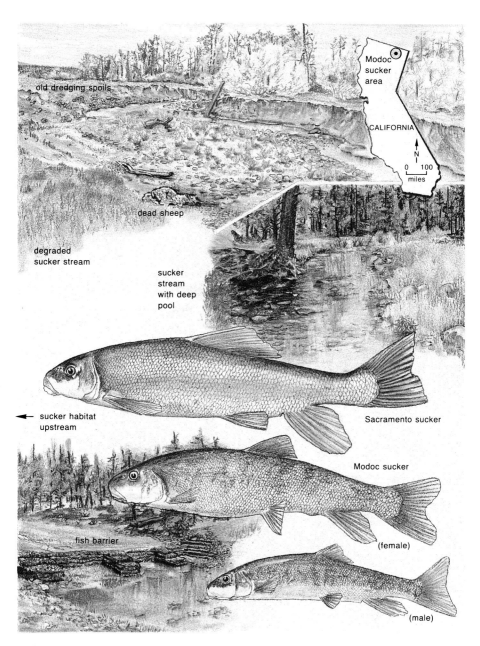

Plate 7. The Modoc sucker is threatened with extinction because of a combination of factors: overgrazing and channelization of its streams, predation from introduced trout, dams and diversions, and hybridizing with invading Sacramento suckers. It has been saved through a program of stream restoration and construction of barriers that prevent invasion by nonnative fishes.

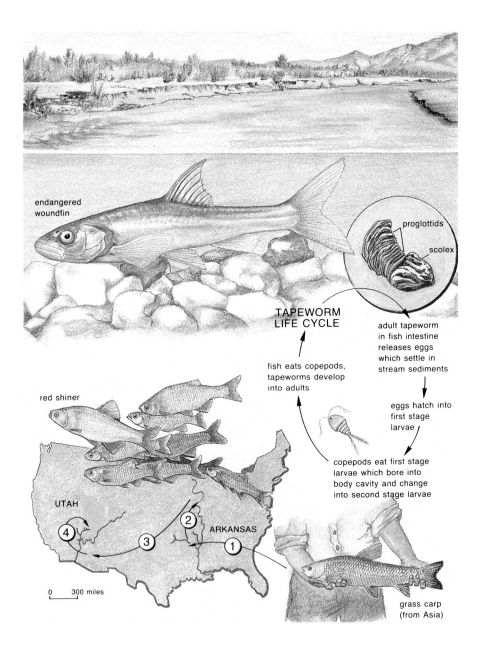

Plate 8. A factor contributing to the decline of the woundfin minnow of Utah is their infestation by an Asiatic tapeworm. The tapeworm was brought in by the grass carp when it was introduced into Arkansas from Asia (1). Grass carp then colonized the Mississippi River and spread the tapeworm to red shiners (2). Infested shiners were brought to Utah as bait fish (3), were released, and became established in the Virgin River (4), where they infected woundfin with tapeworms. This consequence of the introduction of grass carp could not have been predicted.

Figure 15-1. Since 1885, when Samuel Marsden Brookes painted these chinook salmon in California, chinook salmon populations have declined enormously despite their high economic value. Today, intense conservation efforts are needed not only to protect "valuable" fish like salmon, but a myriad of other species with little obvious value to humans. (*Steelhead Salmon*, 40×30 in., oil on wood panel, courtesy The Oakland Museum, Oakland, California)

TABLE I.   NUMBERS OF FISH SPECIES THAT ARE
EXTINCT OR IN DANGER OF GOING EXTINCT SOON
(ENDANGERED), LIKELY TO BECOME ENDANGERED
(THREATENED), OR WITH SEVERELY DECLINING
POPULATIONS (SPECIAL CONCERN) IN SELECTED
COUNTRIES OR REGIONS OF THE WORLD.

Region	Endangered	Threatened	Special Concern	% of Total Fauna
North America	103	114	147	31
California	30	21	22	63
Arkansas	11	16	15	21
Europe	21	40	19	42
Iran	1	15	19	22
South Africa	4	31	27	63
Sri Lanka	2	7	2	17
Australia	6	6	37	26
Costa Rica	1	1	12	9

SOURCE: Data from Moyle and Leidy, 1992 (see chapter 16).

as we do. This attitude, of course, puts an enormous burden of respon-
sibility on humanity, as we are the first species to be able to consciously
affect the fate of most other species on this planet. The High Rock
Spring tui chub went extinct because no one cared enough to prevent it
from doing so. The government issued the proper permits for the introduction of other fishes (for aquaculture purposes) and the rancher allowed the exotic fishes to take over the spring. It is likely that other
organisms have gone extinct along with the tui chub, perhaps a snail
species or two, or some other obscure invertebrate species, all unknown
and forever unknowable. Such extinctions were not necessary and were
wrong, even if legally permitted. This ethical position is developed in
detail by such writers as Aldo Leopold, B. G. Norton, J. B. Callicott,
and R. F. Nash.

Aesthetically, I look at each species in the way we are taught to ad-
mire the work of great artists such as Leonardo, Rembrandt, or Picasso,
as an irreplaceable thing of beauty. The beauty stems not so much from
external appearance as from appreciation of how individual members
of each species live and function as an integrated part of their environment. Each species has a marvelously adapted set of characteristics that
have allowed it to make a long journey through time, changing as the

environment changes. I know little about the High Rock Spring tui chub and am poorer for not knowing how this fish managed to persist in a small, isolated desert spring, miles from other fish populations. It is likely that thousands of years ago a lake occupied the region around the spring, as most other tui chubs are lake-dwelling fishes. Somehow, when the lake dried up, the last of the chubs survived in the spring, whereas other lake fishes became extinct. The chubs adapted to this much more limited environment but were presumably ready to burst out again if a wetter climate returned, and with it the lake. Yet we were able to destroy this miniature masterpiece of evolution in a few months, without thought or shame.

Practically, there are many reasons to save even obscure species like the High Rock Springs tui chub, although they tend to be future oriented values that are easily overcome by short-term financial gains an extinction might cause. Tui chubs, for example, were prized as food by Native Americans but have largely been rejected as such by our modern culture, because they are bony and are not caught easily by anglers. Yet cultural values change and fishes like tui chubs could become a valuable resource, especially if they can live under crowded conditions and feed low on the food chain (like the High Rock Springs tui chub). Native fishes are also good indicators of environmental quality; they can serve as early warnings of large-scale phenomena such as acid rain. For example, the tiny Delta smelt is a great indicator of conditions in the upper Sacramento–San Joaquin estuary because it has a one-year life cycle; if conditions are poor for it, and other fishes, its population the following year is small; if they are good, the population quickly grows. The fact that this species is endangered is an indication that the entire biota of the upper estuary is in trouble. When action is taken to save the Delta smelt, this same action will increase the populations of "important" fish such as chinook salmon, striped bass, and American shad. Unfortunately, the short-term gains of an aquaculture operation doomed the High Rock Springs tui chub. The short-term value of the water (for irrigated agriculture) in which the Delta smelt lives may doom it as well, along with many other fishes that live with it.

WHY FISH ARE IN TROUBLE

I first began to understand why fish are often in trouble when I was a seasonal worker for the Montana Department of Fish and Game, in the summers of 1961 and 1962. Part of my time was spent on crews surveying cutthroat trout populations in pristine streams near Glacier National Park, so the contrast with our other duties was great. These alternate duties consisted of measuring the damage done by road construction to streams. We found that the channels of some streams were tightly confined between a highway on one bank and a railroad on the other. Substantial numbers of fish were found only in places where such streams were allowed to break from their confinement and meander, creating deep pools. We also sampled trout populations in streams that were damaged by logging, by sewage pollution from small towns, and by heavy grazing; we surveyed streams that had become nearly fishless from the effects of mining and smelting of copper ore. We set nets in reservoirs behind huge dams to try to find out why fishing in them was so poor, compared to that in the streams they replaced.

One week I spent in Bitterroot National Forest observing attempts by the U.S. Forest Service to control an outbreak of a forest insect pest using massive applications of the pesticide, DDT. I watched as a converted B-17 bomber flew low over forest, lake, and stream, followed by long trails of spray that settled slowly among the trees. After an area had been sprayed, I sampled the streams with a piece of screen between two poles. I was supposed to hold it in the water for two minutes and then count the number of aquatic insects that collected on it, killed by the insecticide. Often the screen would become too clogged with dead insects for me to hold it for the required length of time. Direct kills of trout from the spraying, however, were not observed. In subsequent years fishing was poor in the streams, however, as trout either starved to death, died undetected later on, or failed to reproduce, all the result of delayed effects of the pesticide. My supervisor, Arthur Whitney, suggested that I read Rachel Carson's 1957 book, *Silent Spring*, to understand what was happening. I did.

The many abuses of streams I observed in Montana suggest three generalities that explain why fish are increasingly in trouble:

1.   Water flows downhill and accumulates in the lowest places as streams and lakes. Therefore abuses of the land will be magnified in the waters.

2.   Most human activities take place close to streams, lakes, and estuaries and alter them as a consequence.

3.   Human abuse of aquatic systems frequently exceeds the capacity of the systems to absorb damage.

Today there is no lake or stream and no part of the ocean that has not been touched by human activity. Remote lakes receive pesticides and acidity in the rain that gives them water. Mountain streams are diverted and the water sent hundreds of miles through pipes to our cities. Rivers and lakes receive enormous quantities of practically every known form of human waste, and their shores and banks are altered for our convenience. Species of fish deemed desirable by humans or adapted to the altered environments have been spread willy-nilly over the planet, mostly in fresh water but increasingly in marine environments as well. The surprising thing about all this abuse is how many native fishes are still around and how many habitats still seem to have whole faunas in one piece. This means conservation of many aquatic habitats is still possible and restoration of degraded habitats is feasible. In others, restoration is impossible because so many native species have gone extinct. We can use these irreversibly altered situations as bad examples to learn from and as opportunities for creating new ecosystems, incorporating as many of the native elements as possible.

The rest of this chapter is devoted to examples, good and bad, that should help give some reality to the above generalizations. The examples are also the subject of the color illustrations in the center of this book.

THE FISH THAT ATE LAKE VICTORIA

Lake Victoria is one of the great rift lakes in East Africa. Each lake is enormous in size, ancient in age, and supports (or used to support) an incredible array of endemic fishes. Most of the native fishes of these lakes are haplochromine cichlids, small, spiny-rayed, deep-bodied fishes that are among the most specialized of all vertebrates in their ways of making a living. For example, there are species that live by:

1. Plucking the scales from other species.

2. Ramming mouthbrooding fishes to force them to disgorge their young, which are then eaten.

3. Scraping algae from the leaves of plants.

4. Scraping diatoms from rocks.

5. Plucking tiny invertebrates from plants with forceps-like teeth.

6. Scooping up big mouthfuls of sand and then separating the sand and the invertebrates that live in it.

7.   Nibbling on hippopotamus dung.

8.   Hiding inside snail shells.

9.   Playing dead and then snapping up small fishes that come to feed on the "carcass."

Most of these species are undescribed and we are just beginning to understand the ecology and evolution of the two hundred to five hundred species that live in each lake. We do know that the amazing degree of specialization permits enormous numbers of fish to use efficiently the resources available and to support large subsistence fisheries by local peoples. Increasingly, the bright multicolored cichlids with their bizarre specializations are being prized by aquarists and by tourists who come to snorkel over the rocky reefs where fish concentrate. Unfortunately, the hundreds of species that existed in Lake Victoria are now either extinct or greatly diminished in numbers, devoured by introduced Nile perch.

The Nile perch is a fish very much favored by Westerners because it is a good game fish that grows quite large (up to 2 meters long and 200 kilograms in mass) and can be reduced to large tasty fillets, with few bones. Western fisheries managers suggested that it be introduced into Lake Victoria to improve the fisheries, because the small cichlids that were caught by the local people were regarded as inferior fish, best used as food for bigger fish. The perch were first introduced experimentally into a small lake nearby, where they proceeded to wipe out the local flock of cichlid species. Before this experiment was evaluated, however, the perch mysteriously appeared in Lake Victoria and gradually spread throughout the lake, feeding on the abundant haplochromines. The small cichlids are now largely gone from the lake and the perch are feeding mainly on their own young and a native shrimp species. The lake ecosystem seems to have been transformed from an incredibly complex and balanced system to a comparatively simple one, probably very unstable.

From a narrow fisheries perspective, the introduction of Nile perch has been regarded as a success. In the 1950s, the fishery in one bay of Lake Victoria landed 650,000 haplochromines and 300,000 individuals of nine other species; in 1982, the fishery brought in 250,000 Nile perch, no haplochromines, and very little else. The fishery in the lake has thus switched to perch and an export market for perch has developed, which brings more cash into the local economy. Curiously, some fishery experts argue

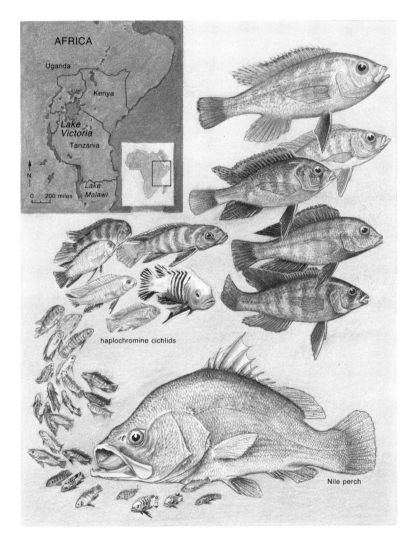

Figure 15-2. The introduction of predatory Nile perch into Lake Victoria in east Africa has resulted in the probable extinction of about two hundred species of small cichlids found only in the lake. These fish were valuable as food fishes, as aquarium fishes, and as marvels of evolution.

that the "undesirable" haplochromines were being overfished at the time of the introduction and might have collapsed anyway!

Despite the thriving perch fishery, the long-term prognosis for the local economy is not good, as total fish production in the lake has become less and it is highly likely the Nile perch fishery will collapse from overfishing. Because the perch are large and oily, they cannot be sun-dried the way the small haplochromines were, to preserve them. Instead, they must be cooked in their own oil, a process that requires wood to keep the fires going. Deforestation of islands and lake shores is occurring as a consequence, with many unknown consequences to local people and other life. One can hope (probably unrealistically) that local fisheries managers will learn from this experience and not introduce Nile perch into the other Great Lakes of Africa. The Lake Victoria disaster has been used already to justify banning the importation of Nile perch into Australia and to initiate a careful program of evaluating introductions in New Guinea.

DISAPPEARING FISHES OF THE GRAND CANYON

The Colorado River is the mighty stream that carved the Grand Canyon, one of the most impressive features on the Earth's surface. The erosive powers of the river made it brown with silt, and huge, violent floods were a regular feature of its history. This turbulent habitat was nonetheless home to a unique collection of fishes, evolved to thrive under conditions favored by few other fishes. Today virtually the entire fauna is endangered, because the river has been tamed to provide water and power for farmland and cities.

The Colorado River drains 600,000 square kilometers of the western United States, including the western slope of the Rocky Mountains. In all this area, there are only thirty-two species of native fish, 75 percent of them found nowhere else. Most remarkable of these fish are the four species that live in the main river, including the Grand Canyon reaches. They are large and very streamlined to withstand the strong currents. Their heads are small and depressed with small eyes, whereas their backs are humped, often with sharp keels, and their elongate bodies narrow down to thin caudal peduncles. Their fins are large, with stout leading rays, and their skin is tough and leathery, with deeply imbedded scales. The largest of these big river fish is the Colorado squawfish, a predator on the other fishes, which reaches 1 to 1.5 meters long and makes long migrations to favored spawning areas. A common prey of the squawfish is the razorback sucker, with its sharply keeled back (giving it a triangular cross-section) and large pectoral fins, adaptations that allow it to feed on the bottom in fast water. These fish can apparently live forty to fifty years and reach 75 centimeters long. Even more bizarre in appearance, if that were possible, are the descriptively named bonytail and humpback chub, which feed up in the water column. Found with these four species in the Grand Canyon, but usually in less severe habitats, are four other native fishes: speckled dace, roundtail chub, flannelmouth sucker, and bluehead sucker.

The four highly specialized species are approaching extinction today, with just a few remnant populations in the wild. Their best hope probably lies in captive rearing programs. Even in the Grand Canyon, a national park since 1919, these fish have largely disappeared. The wild waters where rafters go to seek thrills in a "pristine" wilderness are increasingly unsuitable for the native fishes. In spite of this, at least some of the fishes, especially humpback chub, are still hanging on in the canyon waters. This is a considerably better situation than the one that exists in the lowermost three hundred miles or so of river, including the section that separates California and Arizona. This reach has the distinction of being the only large river in the world that contains **only** introduced species of fish.

The native fishes have disappeared because the Colorado River is probably the most developed river in the world. Less than 1 percent of its water now reaches its mouth on the Gulf of California. From headwaters to mouth, the river is basically a series of reservoirs separated by sections of regulated stream. The reservoirs were created by some of the biggest dams in the world, triumphs of modern technology. The water is diverted for irrigation and domestic use and generates electricity in the process. Return flows to the river from agricultural areas are polluted with wastes of all kinds. Below the large dams on the main river, however, the water is mostly clear and cold because the sediments have settled out in the reservoirs and the cold bottom water sent downstream. This water supports spectacular trout fisheries but few native fishes can stand its coldness and clarity. The reservoirs also support few native fishes; razorback suckers and bonytail can live in them, but are unable to reproduce and so die in twenty to fifty years without descendants. The giant squawfish also have problems reproducing when dams block their way to traditional spawning grounds. Until recently, few people really cared much about these fish, which were dismissed as "trash fish" or "rough fish." In fact, in 1962, 700 kilometers of the Green River, a major tributary, were poisoned by fisheries management agencies in part to get rid of the native fishes so they would not interfere with fisheries in the newly built Flaming Gorge Reservoir.

Despite all these problems, it is likely that the native fishes could have continued to thrive in the Colorado River system if other fishes had not been introduced into it. Native fishes can live in the reservoirs and in the intervening stretches of river, except the cold, clear ones. They have proven to be easy to breed and to raise in captivity. In short, as Colorado River fish expert W. L. Minckley of Arizona State University points out, they are very adaptable. Their problem is their inability to reproduce themselves, and the reason for this seems to be the introduced fishes. Swarms of nonnative minnows and carp suck up the newly spawned eggs of native fishes or devour the larvae that manage to swim up into the water column. Those few that survive this intense predation on eggs and larvae must survive pursuits by voracious predators such as largemouth bass, smallmouth bass, striped bass, channel catfish, and flathead catfish. Because they grow rapidly, in a year or two the native fishes can grow big enough to avoid such predation, provided they do not succumb to some disease or parasite the exotic fishes brought in with them. Colorado squawfish may actually turn the tables on the introduced fishes and prey on them, but even this does not always work out. Dead squawfish have been found containing catfish, spines erect, jammed in their throats. Overall, the result is a downward spiral in the populations of the native fishes, leading to eventual extinction.

What is being done? Despite the extreme difficulty of working with rare fish in a big river, the native fishes have been, and are being, intensively studied to try to find ways to preserve them. Proposals are being debated to modify flow and temperature regimes in the more natural reaches of river remaining, although it is difficult to enhance habitat for native fishes without enhancing it for nonnatives as well. In the short run, rearing the fish in hatcheries and planting them in the wild at fairly large sizes seems to be the most feasible thing to do. This is being tried at the present time. In the long run, such programs tend to select for fish that are more suitable for hatcheries than for the rivers. In addition, hatchery programs are expensive and can be eliminated by budget cuts, if the political winds favoring endangered species change. Thus finding ways of allowing the native fishes to perpetuate themselves is of para-mount importance.

CUI-UI AND CUTTHROATS IN
THE DESERT

Ten thousand years ago, much of northwestern Nevada was covered by Lake Lahontan. Today most of the old lake bottom is dry desert, home to rabbitbrush, kangaroo rats, and coyotes. The deepest part of ancient Lake Lahontan still persists, however, as Pyramid Lake, an immense blue-green body of water that shimmers in the desert heat, a striking contrast to the bare, brown mountains on all sides. Living in this pro-ductive lake are the remnants of Lake Lahontan's fishes: shoals of tui chubs, redside shiners, Tahoe suckers, and cui-ui, all preyed upon by huge Lahontan cutthroat trout, as well as by flocks of white pelicans, cormorants, and other fish-eating birds. The most distinctive of the fishes is the cui-ui (pronounced quee-wee), a large-headed, heavy-bodied fish that lives fifty or more years while feeding on tiny zooplankton. Considered ugly and inedible by European settlers in the region, it was (and still is) held sacred by the Paiute Indians who prized it for its size, abundance, and reliability as a food source.

White settlers quickly discovered that Pyramid Lake cutthroat trout were very much to their liking and by the late 1800s the trout was being fished commercially to feed railroad workers, miners, and loggers in the nearby Sierra Nevada. Probably around 450,000 kilograms (1 million pounds) was removed from the lake annually when the fishery was at its peak. Later a sport fishery developed, and trout as large as 18.6 kilograms were caught, the largest cutthroats on record. Despite its value, by 1943 the Pyramid Lake cutthroat trout was extinct, the result of overharvest combined with the diversion of water from its sole spawning stream, the Truckee River. The diversion, Derby Dam, was constructed in 1905 to provide water for growing alfalfa under the des-ert sun. This dam was also responsible for the decline of the cui-ui.

The reason the construction of Derby Dam had such serious conse-quences is that it diverted half the flow of the Truckee River, the only permanent source of water for Pyramid Lake. The lake level dropped drastically and a shallow delta formed at the mouth of the river with channels too shallow and braided for easy passage of fish. Those fish that managed to find a swimmable channel were easy prey for predatory birds. In wet years, a few trout and cui-ui would make it up into the river, but the trout found their passage to upstream spawning grounds blocked by the dam. The cui-ui were able to spawn below the dam and thus could persist, but the trout became extinct. The cui-ui also had the advantage of living much longer than the trout. When Gary Scoppettone of the U.S. Fish and Wildlife Service studied cui-ui in 1984, he found that almost all the fish he examined were either fifteen or thirty-four years old, reflecting successful spawning in two wet years. In the long run, however, even long life will not protect the cui-ui, as the lake is getting progressively smaller and saltier, as water evaporates but is not replaced by inflow from the river. Since 1905, the lake has dropped over 25 meters in water level, creating an immense bathtub ring of barren, rocky shoreline around it.

Surprisingly, there is hope for the survival of the cui-ui and the Pyramid Lake ecosystem. The first step toward recovery was the restoration of cutthroat trout to the lake, which began in the 1950s. These trout are of a different strain than the native trout and do not seem to grow as large, but they do function as a top predator again and support a valuable sport fishery. All of the trout are reared, however, in hatcheries and planted in the lake. When ready to spawn, many of them are attracted to the effluent of the Sutcliffe hatchery on the shores of the lake, rather than migrating up the Truckee River. The Paiute Indians have also established a hatchery for cui-ui, although the success of the program is limited. A more spectacular development to save the fish was the construction, in 1976, of Marble Bluff Dam, just above the Truckee River Delta. This dam diverts water into a 4.8-kilometer-long fishway that lures cui-ui and trout around the delta and into the river above the dam. If a fish misses the fishway and happens to make it through the delta to the bottom of the 12-meter-high dam, it is shunted into an elevator, which lifts it over. Unfortunately, these elaborate developments have had only limited success; most fish seem unable to find either the ladder or the elevator. There has been successful spawning of cui-ui in recent years, although this can be attributed as much to high flows in the river from exceptionally wet years, as to anything else. Eventually, however, the combination of the Marble Bluff facility and the hatcheries should maintain trout and cui-ui populations in the lake, especially if less water is diverted from the Truckee River. The latter may occur as water rights are purchased from farmers and as the soils of the desert farms now watered by the Truckee become too saline even to grow alfalfa. Countering all this, unfortunately, is the ever-growing thirst of Reno, Nevada, through which the Truckee flows, past casinos and hotels. Still, the cui-ui seems to be a remarkably persistent fish. It now not only has Paiute Indians demanding its protection but also descendants of the immigrants who once despised it.

NEW FISH COMMUNITIES FOR OLD: THE GREAT LAKES

The Great Lakes of North America are so immense that it is difficult to comprehend how much their fish faunas have changed through human influence. The changes began with overfishing and alteration of spawning streams by logging and agriculture and were accelerated by repeated invasions of nonnative species and by pollution from cities and factories. Five native species have become extinct: longjaw cisco, deepwater cisco, Lake Ontario kiyi, blackfin cisco, and blue pike, whereas five others are considered to be threatened or endangered: lake sturgeon, bloater, shortnose cisco, and shortjaw cisco. All these species once supported large commercial fisheries. The ciscoes were a complex of closely related species, each of which used a different, fairly narrow depth zone in the lakes. Each zone was defined (as far as the fish were concerned) by temperature, pressure, and food organisms. Although the upper Great Lakes (Michigan, Huron, Superior) are coldwater lakes that are not particularly productive, the specializations of the ciscoes seemed to have maximized the conversion of what production there was into fish, and, for a short while, into fisheries for species prized for their delicate flavor. One of the memories of my childhood is picnicking along the north shore of Lake Superior, with sweet, smoked cisco as the center of the meal, a scarce and expensive delicacy even then. I also remember eating fresh rainbow smelt, which were cheap and abundant.

The abundance of smelt and the scarceness of cisco are tied to each other, as the smelt was introduced into the lakes in the 1920s, where it fed on the same zooplankton eaten by some of the ciscoes and possibly ate cisco eggs and larvae as well. In the 1930s an even more harmful invader appeared, the sea lamprey. Lampreys are voracious predators that latch onto the sides of large fish, rasp a hole in them, and suck out body fluids. Under normal circumstances, lampreys do not do much harm to the populations of their prey; some individual fish may even survive repeated lamprey attacks. Lamprey populations may also have been kept in check by predators of their own. However, in the Great Lakes, the lampreys found prey unusually vulnerable to their attacks as well as spawning streams that allowed them to reproduce in huge numbers. They attacked large individuals of all species, devastating the remaining populations of lake trout and ciscoes, and depleting populations of other fishes not fished commercially, such as suckers and burbot. The final blow to the ciscoes, however, was probably the invasion of the alewife, a herring-like plankton feeder, in the 1950s. In the absence of predatory fishes and with zooplankton abundant, the alewife populations boomed, forming immense schools that scooped up the plankton at all depths, eliminating the food supply for most of the ciscoes. The alewives were so abundant that occasionally they would die by the millions from natural environmental changes to which they were not adapted. Rotting alewives would then litter beaches and get sucked into city water supplies. The clear water of Lake Michigan became cloudy with algae, formerly kept in control by the zooplankton.

Solving these problems has taken an enormous international research and management program, which is still ongoing. Some of the major results of these efforts have been the following:

1.   The amount of pollutants of all types being dumped into the lakes has been greatly reduced.

2.   Lamprey numbers have been controlled, at least temporarily, by trapping and electrocuting adult lampreys on their way upstream to spawn and by killing their larvae in streams using lamprey-specific poisons.

3.   Hatcheries have been built or expanded to raise lake trout, the main native predatory fish, nearly wiped out by lamprey.

4.   Steelhead rainbow trout, chinook salmon, coho salmon, and pink salmon were introduced into the upper lakes starting in the late 1960s to prey on alewife. The populations are maintained through a combination of hatchery production and natural spawning.

The introduction of steelhead and salmon has been a spectacular success. Feeding on alewives, they developed large populations of large fish and attracted hordes of anglers to the lakes, even though health authorities warned that the salmon were so contaminated with pesticides and other pollutants that they were not safe to eat. By the late 1970s the alewife populations had dropped to low levels and stayed there, thanks to the combination of fish predation and cold winters (which alewife have a hard time surviving). Today the salmon fishery has tapered off and some anglers are demanding that something be done about restoring the "endangered" alewife! The waters of Lake Michigan are clear again, however.

It is clear that a new ecosystem has developed in the upper Great Lakes, one that is fairly unstable because it is maintained partly by artificial means (lamprey control and fish hatcheries) and partly because new players are constantly being added. In recent years three species have been introduced that have the potential for causing further major changes in the lakes; all were introduced accidentally from Europe in the ballast water of ships. The spiny waterflea is a predatory zooplankton species, the zebra mussel is a filter-feeding clam that can coat the bottom in places, and the ruffe is a small perch that may prey on the young of other fishes. Measures are now being taken to prevent future ballast water introductions, but it is obvious that the Great Lakes fish communities will continue to change and that the complex, delicately balanced cisco–lake trout community is gone forever.

OWENS PUPFISH: RESCUED FROM THE BRINK OF EXTINCTION

In August 1969, a student employee of the California Department of Fish and Game noticed that a small, reed-choked desert pond was drying up. Alarmed, he reported it to his boss, biologist Phil Pister. He knew his report would be taken seriously because Pister was one of the few agency biologists of that time who actually thought obscure inedible fish were important—and the pond was the last remaining habitat of the Owens pupfish. Quickly, a rescue effort was mounted, and eight hundred pupfish were captured and placed in cages in the nearby channel. In a few hours, the Owens pupfish would have been extinct.

The Owens pupfish is one of a number of species of pupfish that inhabit isolated desert springs and salty streams throughout the North American Southwest. They are small (3–5 centimeters) deep-bodied fish that are incredibly tolerant of the extremes in temperature and salinity often encountered in desert waters. They were named pupfish because the first people to keep them in aquaria observed "playful" behavior by the brilliant, iridescent blue males. What was interpreted as play was in fact the deadly serious matter of courting the mottled brown females. Although the busy behavior of the colorful males make pupfish interesting to keep in aquaria, there are many species and subspecies of pupfish that look pretty much alike; often only an ichthyologist can tell the difference. They therefore are classic examples of fish that seem to have little justification for protection by government agencies. In a water-short region, "practical" people may become infuriated when water is set aside for tiny fish that cannot even be eaten, especially when the water seems to be needed for drinking or bathing, for watering livestock or for growing crops.

Pister had once proudly considered himself one of those practical people and was devoting his career to improving trout fishing in local streams and reservoirs, often by planting fish. However, in 1964 Pister had been dragged away from his work by two eminent ichthyologists, Robert R. Miller and Carl L. Hubbs, to look for pupfish. The Owens pupfish had last been seen in 1956 and was thought possibly to be extinct. A careful search, however, located one remaining population. The enthusiastic dedication of Miller and Hubbs to desert fishes so impressed Pister that he became an advocate of all desert fishes and of the unique environments in which the fishes lived. In his own words: "Now with the blinders off, I could view the native fishes as part of a biological scheme and balance created with infinite precision by a power well beyond man's absolute comprehension. No longer would I smugly attempt to rewrite Beethoven in a different key or touch up Rembrandt to fit my individual and immature taste" (E. P. Pister, 1985, "Desert Pupfishes: Reflections on Reality, Desirability and Conscience," *Environmental Biology of Fishes* 12:5–6).

In the following years, the three biologists decided that the only way to save Owens pupfish in the long run was to create a refuge for it, along with the three other species endemic to the Owens Valley: Owens sucker, Owens tui chub, and Owens speckled dace. Fish Slough, north of Bishop, was chosen as the most appropriate site, as it was a remnant of the historic habitat for all the species. Originally, the four fishes had been abundant throughout the valley, but the Owens River was diverted first for farming and then in the early 1900s to provide water for the city of Los Angeles, several hundred miles away. What water remained in the Owens River or in springs and sloughs was largely contaminated with introduced fishes, especially largemouth bass and mosquitofish. Setting up the Owens Valley Native Fish Sanctuary was not easy, requiring funds, permits, and cooperation from a host of private, local, state, and federal agencies. It also required eradicating the nonnative fishes already present and constructing a barrier so they could not reinvade from downstream. The sanctuary was completed in early 1970 and the native fishes were reintroduced successfully.

By 1990, six other refuges had also been established for Owens pup-fish. However, it is still listed as endangered because all the refuges are small and none is secure. Largemouth bass and other fishes are contin-ually being introduced by unsympathetic people into the refuges, or ar-tificial barriers to invasion fail, in one case because of an earthquake. Thus, keeping Owens pupfish and other native species (including plants and invertebrates) from going extinct requires constant vigilance. Al-though skirmishes to save the Owens pupfish continue, they are now part of a much larger campaign to save desert fishes. Pister went on to form the Desert Fishes Council in 1969. This is an organization of sev-eral hundred native fish enthusiasts who have been instrumental in sav-ing desert fishes in both the United States and Mexico. In the case of the Devil's Hole pupfish, council members had to go all the way to the Supreme Court to save it from extinction.

SNAIL DARTER VERSUS TELLICO DAM

This is the story of how a tiny fish almost stopped the construction of a mighty but unnecessary dam. The battle between two vastly unequal forces wound up being a major test of the federal Endangered Species Act of 1973 (ESA). The Act was upheld but the fish lost anyway.

The snail darter is one of over one hundred species of small, colorful perches that inhabit the streams of eastern North America. It was dis-covered in 1973 by David Etnier, an ichthyologist at the University of Tennessee, who described it as a new species. Unfortunately, the only place he found it living was in about 25 kilometers of the Little Ten-nessee River in Tennessee, one of the last free-flowing sections of river in a region littered with Tennessee Valley Authority dams. Construction of Tellico Dam, which would flood the reach, was already under way despite protests of environmentalists who wanted to save the river and of Cherokee Indians who wanted to save historic and sacred village sites. In 1975, the snail darter was listed as endangered and the Little Tennessee River was listed as the critical habitat for it. In 1978, the Supreme Court upheld a lower court decision that the ESA clearly in-dicated that the dam should not be completed, even though $100 mil-lion had been spent building it. As a result, Congress amended the ESA to establish the Endangered Species Committee (commonly known as "the God Committee"), a group of cabinet secretaries and agency heads who could overrule endangered species decisions made under the act. The committee promptly ruled in favor of the darter, pointing out that the dam was not really needed. In 1979, however, Congress passed a bill funding water and energy projects all over the country that included a spurious and clandestine exemption from ESA for Tellico Dam. President Carter reluctantly signed the bill, the dam was completed, and the river became a reservoir.

Meanwhile, the darter survived in populations established by moving fish to a nearby stream. Then in 1980, another population was discovered in South Chickamanga Creek in Tennessee and Georgia. Further searching revealed five additional populations, although none is secure because of pollution and other factors. The discoveries did cause the darter to be "down-listed" to threatened status, however. The snail darter controversy demonstrated three important things:

1.   The Endangered Species Act is a strong, well-written act that can protect species, but economic considerations can cause it to be overridden.

2.   Because of the strength of ESA, protecting endangered species is being used as a surrogate for protecting ecosystems. This has the unfortunate consequence of focusing fights on the "insignificant" species rather than on the endangered ecosystems in which they live.

3.   There is a need for systematic biological surveys of our nation's waters, so we really do know the status of all our fishes and other aquatic organisms.

Unfortunately, the controversy also has created a certain amount of timidity among agencies and individuals in charge of listing endangered species. The overriding fear is that if too many species like the snail darter are listed, Congress will rewrite the entire act, taking away protection that now exists for many species. Whether or not this fear is justified remains to be seen.

CONCLUSIONS

1.   The most endangered aquatic ecosystems are lakes, big rivers, and small, isolated waters.

The examples demonstrate that the fish faunas of our biggest and smallest fresh waters are most affected by human activities. Lakes seem to be most susceptible to disruption by introduced species, whereas rivers are dammed, diverted, channelized, polluted, or otherwise modified for our use. Ecosystems in small waters like desert springs can be destroyed in a single day by someone with a bulldozer or a bucket of predatory fish. The examples in this chapter were chosen because they are particularly dramatic, but they are by no means unusual. In the Mississippi River at least one fish species (harelip sucker) has gone extinct and a number of fishes representing ancient evolutionary lines are in serious decline: paddlefish, pallid sturgeon, lake sturgeon, and blue sucker. Clear Lake in California, one of the oldest lakes in North America, has lost two-thirds of its native fish fauna and is now dominated by introduced species. A number of the pale, blind fishes that live in isolated caves and aquifers of North America are gradually disappearing due to pollution of their water, its removal by pumping, and from collecting of fish by aquarists: widemouth blindcat, toothless blindcat, Ozark cavefish, northern cavefish, Alabama cavefish, Muzquiz blindcat, Yucatan cave eel, and Mexican cave cusk-eels.

The fact that certain kinds of aquatic habitats are in the most trouble does not mean we can be complacent about the rest—even marine systems. Estuaries around the globe are in serious trouble. The Colorado River scarcely has an estuary anymore, thus a species that depended on it, the totoaba, is now endangered. The low-salinity parts of the Sacramento–San Joaquin estuary are dominated by introduced species and at least one native species, the delta smelt, is endangered. Gradually the negative effects of the expanding human population are spreading seaward and we may be seeing more endangered marine fishes in the not too distant future.

2. Aquatic species and ecosystems are endangered for multiple reasons.

The examples should make it obvious that species and ecosystems rarely decline from single causes or events, the Nile perch introduction into Lake Victoria being one of the major exceptions. This generality became obvious to *me* when I did my first study of an endangered species, the Modoc sucker. This 15- to 20-centimeter-long fish lives in a few small headwater streams in northeastern California. The banks of these streams had been badly overgrazed by cattle and sheep, so there was little cover from overhanging plants, and the pools the suckers prefer were often filled in with silt. Some stream sections had been channelized for flood control, whereas others had been altered for roads and bridges. Water was being diverted to grow Christmas trees. Natural barriers had been removed so the streams were being invaded by Sacramento suckers, a closely related but abundant species that hybridizes with Modoc suckers. Brown trout had been introduced, which preyed on the suckers. A major dam was proposed that would flood some Modoc sucker habitat. In fact, I was surprised the suckers were still surviving at all. Not surprisingly, saving the Modoc sucker has required multiple actions by the California Department of Fish and Game and the U.S. Forest Service. Barriers have been built to prevent further invasions of nonnative fishes. Cattle have been excluded from the most critical sections of the stream to allow the riparian vegetation to grow. A variety of stream restoration techniques have been applied to create more deep pools. Some streams have been treated with fish poisons to get rid of exotic species and the suckers (and other native fishes) have been reintroduced. Despite these efforts, the sucker is still endangered because long-term degradation of its streams has reduced the ability of the streams to respond to management efforts, especially during periods of natural drought.

3.   Introduced species are a growing problem.

Most of the examples of endangered species and ecosystems I have mentioned involve introduced species. All too often they are the final blow that disrupts a system already disturbed by human influences and which drives native species to extinction. Many state, federal, and international agencies formerly promoted introducing species as a way to "improve" aquatic ecosystems. Now these same agencies usually oppose new introductions or require careful evaluations before any introductions are made. Unfortunately, most introductions today are made unofficially. With modern transportation systems and fish-keeping technology, it is all too easy for anyone who wants to introduce fish into any body of water to do so. Game fishes are constantly being moved around in this way, often carried hundreds of miles by enterprising anglers. Fish also get introduced by "accident," when anglers release unwanted bait fish at the end of a day of fishing, when fish escape from fish farms or ponds, when aquarists dump fish they are tired of into the nearest lake or stream, or when a ship's ballast water containing fish and invertebrates is pumped overboard to make room for cargo.

Introductions may have many unexpected consequences. Some introduced species, such as Nile perch in Lake Victoria, have immediate and dramatic effects on the native fauna. Others are more subtle and completely unanticipated. For example, the grass carp from Asia was introduced into lakes in Arkansas, starting in 1968, by the Arkansas Game and Fish Commission. It soon spread to the Mississippi River, where it became widely distributed. The grass carp carried with it from Asia a parasite, a tapeworm. The tapeworm was soon infesting other fishes, including red shiner, a favorite bait fish. Anglers or bait dealers introduced infested red shiners to the Colorado River, and by 1984 the shiners had invaded the Virgin River, a Utah tributary to the Colorado. The tapeworm then infected woundfin, a small native minnow already endangered because of dams and diversions on the river. The high rate of tapeworm infestations in woundfin has been associated with their unexpectedly rapid decline, perhaps because they are weakened by it and less able to compete with red shiners for food and space. Thus the introduction of grass carp in Arkansas ultimately had the effect of further endangering a fish in Utah, a consequence no one could have predicted.

4.   Protecting native fishes protects other native organisms.

This is a truism that usually works, as protecting fishes requires protecting their habitats, which in turn contain many other creatures. For every pupfish species protected, there are probably a dozen species of plants and invertebrates protected, living in the same springs. Obscure as pupfish are, a slogan of "Save the Pupfish" has much more appeal than "Save the Spring Snail!" Increasingly, however, we do need to think in terms of endangered ecosystems; the snail darter controversy showed that species have limited usefulness as surrogates for ecosystems. Moreover, by protecting natural "wild" ecosystems we are protecting their services and values for ourselves.

According to poet Gary Snyder:

> Civilizations east and west have long been on a collision course with wild nature, and now the developed nations in particular have the witless power to destroy not only individual creatures but whole species, whole processes, of the earth. We need a civilization that can live fully and creatively together with wildness. . . . It has always been part of the basic human experience to live in a culture of wilderness. There has been no wilderness without some kind of human presence for several hundred thousand years. Nature is not just a place to visit, it is *home*. (Gary Snyder, *The Practice of the Wild*, 1990 [San Francisco: North Point Press], pp. 6–7)

CHAPTER XVI

Resources for the Aquatic Naturalist

The best tools that a naturalist has are an open mind, patience, and keen senses. Among the most pleasurable things a naturalist can do is to use these tools in a pleasing environment: a rushing stream, a pond, a tide pool, or a quiet bay. There is no question, however, that natural experiences can be enhanced by knowledge gained from previous studies and by the many tools modern technology provides at reasonable cost. The purpose of this chapter is to provide some access to the literature and other tools for the serious naturalist. I will first provide a short discussion of literature relevant to each chapter and then discuss other literature and tools useful to the naturalist. Naturally, the information provided here is incomplete. It simply reflects my own biases about what is important, so should be used mainly as a starting place.

CHAPTER 1: GENERAL REFERENCES ON FISHES

This book is based in good part on the text *Fishes: An Introduction to Ichthyology* by myself and Joseph J. Cech, Jr. (2nd edition, 1988, Prentice-Hall, Englewood Cliffs, N.J.). The chapters there on systematics and ecology provide more detailed discussions of many (but not all) of the topics covered in this book and contain many references to the primary literature. Other North American texts are *Ichthyology* by K. F. Lagler, J. E. Bardach, R. R. Miller, and D. R. M. Passino (1977, Wiley, New York) and Carl Bond's *Biology of Fishes* (1979, W. B. Saun-

Figure 16-1.   An example of gyotaku (fish printing).

ders, Philadelphia). A highly readable, if dated, text is N. B. Marshall's *The Life of Fishes* (1966, Universe Books, New York). Heavily illus-trated, semipopular books on fishes are many, so I will just mention a few that I have found to be handy references: *Living Fishes of the World* by E. S. Herald (1961, Doubleday, Garden City, N.J.), *Freshwater Fishes of the World* by G. Sterba (1962, Vista Books, London), and *The Aquar-ium Encyclopedia* also by Sterba (German edition, 1978, but English translation published in 1983 by MIT Press, Cambridge, Mass.). The latter two books are aimed at aquarists but have fairly comprehensive coverage of fishes in general. Perhaps because the majority of amateur ichthyologists have been aquarists, Tropical Fish Hobbyist (TFH) Pub-lications has long been a leader in publishing books on fishes, with color illustrations, including the reissuing of classic works. TFH books are usually sold in aquarium stores but can also be obtained through TFH Publications, Inc., 211 W. Sylvania Ave., Neptune City, N.J. 07753.

CHAPTERS 2 AND 3: ANATOMY

Fish anatomy is covered in the texts listed above in some detail. Dis-section guides generally are available in university bookstores or through biological supply companies (see below), but usually just cover the dogfish shark and the yellow perch. The most widely used of such dissection manuals are published by the William C. Brown Co., Dubuque, Iowa.

CHAPTER 4: BEHAVIOR

There are numerous recent texts on animal behavior that include many examples from fish. Particularly relevant for natural historians is *An Introduction to Behavioural Ecology* by J. R. Krebs and N. B. Davies (1981, Sinauer Associates, Sunderland, Mass.). Two classic and thoroughly delightful popular books that provide a good idea of how naturalists can approach the study of animal behavior are *Curious Naturalists* by Niko Tinbergen (1958, Basic Books, New York) and *King Solomon's Ring* by Konrad Z. Lorenz (1952, Crowell, New York). Tinbergen and Lorenz shared a Nobel Prize for their pioneering work in animal behavior, including studies of sticklebacks. If you want to get a good sampling of extraordinary ways in which fish reproduce, browse through *Modes of Reproduction in Fishes* by C. Breder and D. E. Rosen (1966) or *Reproduction in Reef Fishes* by R. E. Thresher (1984), both issued by TFH Publications, Neptune City, N.J.

CHAPTER 5: DIVERSITY

The classification system used in this book follows *Fishes of the World*, 2nd edition, by J. S. Nelson (1984, Wiley, New York), which includes brief descriptions of all the fish families of the world. *A List of the Common and Scientific Names of the Fishes of the United States and Canada* (1991) is published by the American Fisheries Society (5410 Grosvenor Lane, Bethesda, Md. 20814-2199) and is revised every ten years. Common names used in this book follow those recommended by the American Fisheries Society.

Freshwater fishes. The best single reference for identifying North American fishes is *A Field Guide to Freshwater Fishes: North America North of Mexico* by L. M. Page and B. M. Burr (1991, Houghton Mifflin Co., Boston) which is part of the Peterson Field Guide Series. An illustrated key to the same fishes is *How to Know the Freshwater Fishes* (3rd edition) by S. Eddy and J. C. Underhill (1978, William C. Brown Co., Dubuque, Iowa). Both books are excellent but are a bit cumbersome to use because they cover so many species. Accounts of these species, emphasizing distribution, are found in one volume in *Atlas of North American Freshwater Fishes*, edited by D. S. Lee et al. (1980, North Carolina State University, Museum of Natural History, Raleigh, N.C.).

Most states or regions of North America have their own guides to the freshwater fishes. They vary tremendously in quality and content so I will list here only some of the more comprehensive ones. If you are fortunate enough to live in Canada or adjacent states, *Freshwater Fishes of Canada* by W. B. Scott and E. J. Crossman (1973) is an information-packed volume available at a reasonable price from Information Canada (Ottawa, K1A 0S9). Thanks to the strong ichthyological traditions in the midwestern United States, a number of excellent volumes are available for that region. *Fishes of Wisconsin* by G. C. Becker (1983, University of Wisconsin Press, Madison) is an extraordinary, if weighty, volume, filled with tidbits of natural history. More succinct, with excellent keys, are *Fishes of Missouri* by W. A. Pfleiger (1975, Missouri Dept. of Conservation, Columbia), *Fishes of Illinois* by P. W. Smith (1979, University of Illinois Press, Champaign), *Fishes of Ohio* by M. B. Trautman (2nd edition, 1981, Ohio State University Press, Columbus), and *Fishes of the Minnesota Region* by G. L. Phillips, W. D. Schmid and J. C. Underhill (1982, University of Minnesota Press, Minneapolis). A book that is incomplete in its coverage of nongame fishes but which has wonderful paintings of many species is *Fishes of the Central United States* by J. R. Tomelleri and M. E. Eberle (1990, University Press of Kansas, Lawrence 66045). For the New England states, try *Fishes of New York* by C. L. Smith (1986, Department of Environmental Conservation, Albany).

For the South, the choices are fewer despite a richer fish fauna, although books on the fishes of Tennessee, Mississippi, and Florida are in progress and should be watched for. The best by far is *Fishes of Arkansas* by H. W. Robison and T. M. Buchanan (1988, University of Arkansas Press, Fayetteville). Other useful books are: *The Fishes of Kentucky* by W. M. Clay (1975, Kentucky Department of Fish and Wildlife Resources, Frankfort), *The Freshwater Fishes of Georgia* by M. D. Dahlberg and D. C. Scott (1971, Georgia Academy of Sciences, Athens), *Freshwater Fishes of Alabama* by W. F. Smith-Vaniz (1968, Auburn University, Auburn), and *Freshwater Fishes of Louisiana* by N. H. Douglas (1974, Claitors Publishing, Baton Rouge).

In the West, most states have books or pamphlets describing their

fish faunas but among the more recent are *Fishes of the Great Basin* by
W. F. Sigler and J. W. Sigler (1987, University of Nevada Press, Reno),
Fishes of Idaho by J. C. Simpson and R. L. Wallace (1978, Idaho Research Foundation, Moscow), *Inland Fishes of Washington* by R. S. Wy-
doski and R. R. Whitney (1979, University of Washington Press, Seat-
tle), and *Inland Fishes of California* by P. B. Moyle (1976, University of
California Press, Berkeley, Los Angeles, London). A field guide to Cal-
ifornia fishes is *Freshwater Fishes of California* by S. M. McGinnis
(1984, University of California Press, Berkeley, Los Angeles, London).
In the Southwest, try R. J. Miller and H. W. Robison's *The Fishes of
Oklahoma* (1973, Oklahoma State University Press, Stillwater), W. L.
Minckley's *Fishes of Arizona* (1973, Arizona Department of Fish &
Game, Phoenix) or J. E. Sublettes, M. D. Hatch, and M. Sublette's
Fishes of New Mexico (1990, University of New Mexico Press, Albu-
querque). The latter book is unique in that it is illustrated with fasci-
nating electron microscope photographs of fish structures.

Marine fishes.   Books on marine fishes are generally fewer and less
comprehensive than books on freshwater fishes. Two of the best are J. L.
Hart's *Pacific Fishes of Canada* (1973) and W. B. and M. G. Scott's *Fishes
of the Atlantic Coast of Canada* (1988). Because fish do not respect political boundaries, both books are very useful in the United States as well.
Both are available from Information Canada (see above). For the Atlantic
Coast, the best field guide is *A Field Guide to Atlantic Coast Fishes* by
C. R. Robins, G. C. Ray, and J. Douglass (1986, Peterson Field Guide Series, Houghton Mifflin Co., Boston). For the Pacific Coast, two excellent
field guides are available: *Guide to the Coastal Marine Fishes of California* by D. J. Miller and R. N. Lea (1972, available from Division of Agricultural Sciences, University of California, 1422 S 10th St., Richmond,
Calif. 94804) and *A Field Guide to Pacific Coast Fishes of North America*
by W. N. Eschmeyer and E. S. Herald (1983, Houghton Mifflin Co.; this
is also one of the Peterson Field Guides). Less comprehensive, but containing excellent underwater photographs of fishes commonly seen by divers is *Pacific Coast Inshore Fishes* by D. W. Gotshall (1989, 3rd edition,
Sea Challengers, Monterey, Calif.). A delightful volume on the biology of
many species is *Probably More Than You Want to Know About the Fishes
of the Pacific Coast* by R. M. Love (1991, Really Big Press, P.O. Box
60123, Santa Barbara, Calif. 93160). For the Gulf of California, see the
discussion under reef fishes.
    A handy book and complete guide for the Gulf of Mexico is *Fishes*

of the Gulf of Mexico by H. D. Hoese and R. H. Moore (1977, Texas A&M Press, College Station), although *Dr. Bob Shipp's Guide to the Fishes of the Gulf of Mexico* by R. L. Shipp (1986, Dauphin Island Sea Laboratory, Dauphin Island, Ala.) is a good guide to the more commonly encountered fishes. For the New England coast, *Saltwater Fishes of Connecticut* by K. S. Thomson et al. (1978, Department of Environmental Protection, Hartford) contains much useful information, as does *Common Marine Fishes of Coastal Massachusetts* by G. Clayton et al. (1979, Bulletin Center, University of Massachusetts, Amherst, 01003).

Sharks. Because sharks excite such popular interest, many books are available on them, a number of them quite good. For identification of sharks, use *The Sharks of North American Waters* by J. I. Castro (1983, Texas A&M Press, College Station). For general information, read *Sharks: An Introduction for the Amateur Naturalist* by S. A. Moss (1984, Prentice-Hall, Englewood Cliffs, N.J.), *The Book of Sharks* by R. Ellis (1975, Grosset and Dunlap, New York; with superb illustrations), or *Sharks in Question* by V. G. Springer and J. P. Gold (1989, Smithsonian Inst. Press, Washington, D.C.).

CHAPTER 6: ECOLOGY

The text of Moyle and Cech (mentioned above) is ecologically oriented. G. V. Nikolsky's 1963 book *The Ecology of Fishes* (reprinted by TFH Publications, Inc.) is badly dated but still makes interesting reading. A more technical book that contains many interesting insights into fish ecology is *Ecology of Teleost Fishes* by R. J. Wooten (1990, Chapman and Hall, New York).

CHAPTERS 7 AND 8: STREAMS

The standard reference on stream ecology, although it is already a bit dated and not particularly easy to read, is H. B. N. Hynes's *The Ecology of Running Waters* (1970, University of Toronto Press). A badly dated but worthwhile account of trout stream biology is P. R. Needham's *Trout Streams* (1969, Holden-Day, San Francisco). For identifying aquatic organisms (in lakes and ponds, as well as streams) a good place to start is *A Guide to the Study of Freshwater Biology* by J. G. and P. R. Needham (1962, Holden-Day, San Francisco). For the serious aquatic entomologist, *An Introduction to Aquatic Insects of North America* by A. W. Mer-

ritt and K. W. Cummins (1984, 2nd edition, Kendall/Hunt Publishing
Co., Dubuque, Iowa) is a must because it contains keys to the families and
genera of aquatic insects. If you live in or visit Minnesota, read *The
Streams and Rivers of Minnesota* by T. F. Waters (1977, University of
Minnesota Press, Minneapolis). For Pennsylvania, try *Rivers of Pennsylvania* by T. Palmer (Pennsylvania State University Press, University Park).
Every state could use such books! In warmwater streams, some of the
most attractive fishes are the darters as is shown in two excellent books
on them: *The American Darters* by R. A. Kuehne and R. W. Barbour
(1983, University of Kentucky Press, Lexington) and *Handbook of Darters* by L. W. Page (1983, TFH Publications, Neptune City, N.J.).

A very useful book for anyone interested in stream ecology and restoration is *Adopting a Stream: A Northwest Handbook* by S. Yates
(1988, Adopt-A-Stream Foundation, University of Washington Press,
Seattle). This provides information and access to tools that any amateur
naturalist can use, including suggestions of activities for schoolchildren.
If you are interested in restoring a local urban stream, a good place to
start is a letter to the Urban Creeks Council, 2530 San Pablo Ave.,
Berkeley, Calif. 94702.

CHAPTER 9: LAKES AND RESERVOIRS

The importance of fish in lake ecosystems is becoming increasingly
widely recognized, but the only recent limnology text that seems to pay
much attention to fish is *Limnology* by C. Goldman and A. Horne
(1983, McGraw-Hill, New York). For methods for the study of lakes,
try O. T. Lind's *Handbook of Common Methods in Limnology* (1974,
C. V. Mosby, St. Louis, Mo.). If your interest is in coldwater lakes, you
should look at *Stillwater Trout*, edited by John Merwin (1980, Doubleday, Garden City, N.Y.), which contains much general information, as
well as instructions on trout fishing.

CHAPTER 10: PONDS

Most state fish and game departments and/or agriculture extension di-
visions of state universities have pamphlets available on setting up and
maintaining farm ponds, oriented to local conditions. A list of these
publications was published in *Fisheries*, the magazine of the American
Fisheries Society ("State Aquaculture Publications" by B. D. McAleer in
vol. 12:1, 1987). For a more general account, consult *Management of*

Artificial Lakes and Ponds by G. W. Bennett (1962, Reinhold, New York). A worthwhile popular account of pond biology is found in *Pond and Brook: A Guide to Nature Study in Freshwater Environments* by M. J. Caduto (1985, Prentice-Hall, Englewood Cliffs, N.J.).

CHAPTER 11: ESTUARIES

The larger estuaries of North America typically have volumes of information written on them, available in agency or university libraries. Descriptive summaries are (or will be) available on them and other coastal habitats in a series published by the U.S. Fish & Wildlife Service's National Coastal Ecosystems Team (write to them at 1010 Gause Blvd., Slidell, La. 70458, for a list). For a technical review of the problems with striped bass, consult the *Transactions of the American Fisheries Society*, 1985, volume 114, number 1, a special issue devoted to the subject. Some of these problems are also covered in the article "Thermal Niches of Striped Bass" by C. C. Coutant (1986, *Scientific American* 255:2). A fascinating book that shows that degraded estuaries can be restored is A. Wheeler's *The Tidal Thames: The History of the River and Its Fishes* (1979, Routledge and Kegan Paul, London). A more technical, but accessible, introduction to estuaries is *The Estuarine Ecosystem* by D. S. McLusky (2nd edition, 1989, Chapman and Hall, New York).

CHAPTER 12: BETWEEN THE TIDES

An excellent introduction to the physical and biological processes that take place in this dynamic region are provided in *At the Sea's Edge* by W. T. Fox (1983, Prentice-Hall), although readers of this book may be inclined to wince at the statement that "vertebrates are notably absent as full-time residents of the (rocky) intertidal zone." This book also discusses coral reefs and other coastal environments and is a good source of references on intertidal ecology. For the Pacific Coast, an excellent book is *Pacific Seashores* by T. Carefoot (1977, University of Washington Press, Seattle) although it almost completely ignores fish.

CHAPTER 13: THE CONTINENTAL SHELF AND BEYOND

Anyone with a serious interest in this vast habitat region will want to learn much about seasick remedies (ask your doctor about scopolamine), oceanography, and marine biology. A good beginning text in oceanography is *Oceanography: A View of the Earth* by M.G. Gross (1982, 3rd edition, Prentice-Hall, Englewood Cliffs, N.J.). Much of the information on the fishes is in the fisheries literature, so books like G. A. Rounsefell's *Ecology, Utilization, and Management of Marine Fisheries* (1975, C. V. Mosby Co., St. Louis, Mo.) or R. J. Browning's *Fisheries of the North Pacific* (1980, Alaska Northwest Publishing Co., Box 4EEE, Anchorage 99509) can be quite informative.

CHAPTER 14: TROPICAL REEFS

The literature on tropical reef fishes is growing rapidly so it is hard to keep up. TFH Publications publishes a number of books on these fishes, with many color plates and texts of varying utility. One of the most fascinating of these books is *Reproduction in Reef Fishes* by R.E. Thresher (1984, address above). For Hawaii and the Indo-Pacific Region, useful guides are R.H. Carcasson's *A Field Guide to the Coral Reef Fishes of the Indian and West Pacific Oceans* (1977, Collins, London), R. and B. Carpenter's *Fish Watching in Hawaii* (1982, Natural World Press, 251 Balwin Ave., San Mateo, Calif. 94401), and J.E. Randall's *Underwater Guide to Hawaiian Reef Fishes* (1981, Treasures of Nature, P. O. Box 195, Kaneohe, Hawaii 96744, printed on waterproof paper). For the Gulf of California, an excellent reference is *Reef Fishes of the Sea of Cortez* by D. A. Thomson, L. T. Findley, and A. N. Kerstich (1979, John Wiley and Sons, New York). More portable is *Gulf of California Fish Watchers Guide* by Thomson and N. McKibbin (1976, Golden Puffer Press, Tucson, Ariz.). For the Caribbean, a good guide is *Handguide to the Coral Reef Fishes of the Caribbean and Adjacent Tropical Waters, including Florida, Bermuda, and the Bahamas* by F. J. Stokes (1980, Lippincott and Crowell, New York). More localized guides are also often available, including plastic cards with common fishes printed on them in color, for comparison while you are in the water. The cards are usually available at local dive shops or from specialized booksellers such as Sea Challengers (4 Sommerset Rise, Monterey, Calif. 93940).

CHAPTER 15: CONSERVATION

Documentation for many of the statements in this chapter can be found in a chapter on conservation of aquatic biodiversity by myself and R. L.

Leidy in *Conservation Biology: The Theory and Practice of Nature Conservation, Preservation, and Management*, edited by P. L. Fiedler and S. Jain (1992, Chapman and Hall, New York). A readable book that discusses endangered fishes is *Vanishing Fishes of North America* by R. D. Ono, J. O. Williams, and A. Wagner (1983, Stone Wall Press, Washington, D.C.). A complete list of fishes in trouble in North America can be found in "Fishes of North America, endangered, threatened, or of special concern: 1989" by J. E. Williams et al. (1989, *Fisheries* 14:2–20). A companion article in the same issue deals with recently extinct fishes of North America. A book that covers all aspects of fish conservation in western North America is *Battle Against Extinction: Native Fish Management in the American West*, edited by W. L. Minckley and J. E. Deacon (1991, University of Arizona Press, Tucson).

If you spend much time as a natural historian, you will inevitably become involved with one or more organizations that have as their goals protecting habitats from unnecessary development, pollution, and other problems. On a national basis, organizations such as the Wilderness Society, Sierra Club, Defenders of Wildlife, and Audubon Society are very effective in lobbying for environmental protection and in educating the public on the issues. Local chapters of such organizations are generally heavily involved in local environmental issues, including the protection of aquatic habitats. One organization whose focus is streams is Friends of the River (1228 N St., Sacramento, Calif. 95814). An extremely effective, if low profile, organization that attempts to protect natural areas through acquisition using private funds is The Nature Conservancy (1800 N. Kent St., Arlington, Va. 22209) which has protected thousands of acres all over the country, including many aquatic habitats. Legal battles for conservation are often fought by groups that specialize in litigation, most prominently Sierra Club Legal Defense Fund, Environmental Defense Fund, and Natural Resources Defense Council. Often leading fights to protect streams, lakes, bays, and estuaries are local angling organizations. An up-to-date directory of conservation organizations as well as of state and federal agencies is published annually by National Wildlife Federation (1412 16th St. N.W., Washington, D.C. 20036).

Some of the most effective environmental groups are those that arise to protect local streams and lakes. I am active, for example, in the Putah Creek Council which is trying to protect and restore our local waterway and fishing hole. If such an organization does not exist in your town, form one! The book by Steve Yates, discussed under streams, provides many ideas on how to adopt a stream and examples of successful programs. Sometimes the achievements of a local group can have implications far beyond the local setting. The Mono Lake Committee has not only managed to achieve major success in saving Mono Lake and its inflowing streams in California, but, by working with the National Audubon Society and a state angling organization (California Trout), has managed to set legal precedents that have statewide and nationwide implications for saving lakes, streams, and other "public trust" resources. Likewise, Save Our Streams (P.O. Box 56, North Fork, Calif. 93643) started out as a group of citizens angry over the proposed hydroelectric development of a favorite local stream and has expanded to work for the protection of small streams all over California.

JOURNALS AND ABSTRACTS

Eventually, a serious naturalist will need to review the scientific literature to find out what is known about a species or body of water that he/she is studying. To do this you will generally need to go to a major research library associated with a university (unless you have a computerized source). Scientific journals likely to be of special interest are: *Copeia* (official publication of the American Society of Ichthyologists and Herpetologists), *Environmental Biology of Fishes, Transactions of the American Fisheries Society, Canadian Journal of Fisheries and Aquatic Sciences*, and *Journal of Fish Biology*. Many states also have academies of science which publish occasional fish papers of local interest in their journals. Many papers on fish, of course, are not published in the above journals, so a literature search must be conducted. The easiest way is to search through abstract volumes, the most comprehensive of which is *Aquatic Sciences and Fisheries Abstracts*. Another useful source is the annual *Current References in Fish Research* available as annual volumes at a low price from Victor Cvarncara (296 E. Hagen Rd., Chippewa Falls, Wisc. 54729).

METHODS FOR THE STUDY OF FISH

Two books that provide descriptions of many of the techniques used in the capture and study of fishes are: *Methods for the Assessment of Fish Production in Freshwater* edited by T. Bagenal (1978, 3rd edition, Blackwell Scientific Publications, Oxford, England) and *Fisheries Techniques* edited by L. A. Nielsen and D. L. Johnson (1983, American Fisheries Society, 5410 Grosvenor Lane, Bethesda, Md. 20814-2199). A companion volume, *Methods for the Study of Fish Biology*, edited by C. Schreck and P. Moyle (1990) is also available from the American Fisheries Society and describes techniques used for working with fish under experimental conditions (including in the field). Angling, of course, is an acceptable way of collecting fish for study (provided the fishing regulations are obeyed). The number of books providing advice on how to fish are a legion but *A Basic Guide to Fishing* by D. Lee (1983, Prentice-Hall, Englewood Cliffs, N.J.) seems to be a good one for beginners.

SOURCES OF EQUIPMENT

Before you undertake a study that involves the capture of large numbers of fish, be sure to check with your local department of fish and game about pertinent regulations and obtain whatever permits are needed. Generally, the least you will need is a valid fishing license. The use of nets, even small seines, is prohibited in many states without a special permit.

Minnow seines, minnow traps, and dipnets are handy items that are often available through sporting good stores, commercial fishing supply stores, or mail-order catalogs of major department stores. A reliable net company that does not seem to object to small orders is Nichols Net and Twine Co., Inc. (RR 3, Bend Road, E. St. Louis, Ill. 62201). A company that produces a variety of aquatic sampling devices (e.g., plankton nets, dipnets, Surber samplers, etc.) is Wildlife Supply Co. (301 Cass St., Saginaw, Mich. 48602). Biological supply companies also handle small nets, as well as a myriad of other types of equipment, from plankton nets to water pollution detection kits. Three of the larger firms are: Wards Natural Science Establishment (P.O. Box 72912, Rochester, N.Y. 14692-9012), Carolina Biological Supply (Burlington, N.C. 27215 and Gladstone, Oreg. 97027), and Turtox/Cambosco (8200 S. Hoyne Ave., Chicago, Ill. 60620). Usually, the biology teachers at your local high school will have copies of these catalogs.

DIVING AND SNORKELING

For snorkeling, you should use a high-quality mask that fits your face well and is easy to clear of water. If you wear corrective lenses, it is possible to buy nonprescription masks with corrective lenses in them or to have prescription lenses made for a mask. The snorkel should be of good quality and fit snugly in the mouth. If you are going to spend much time in cold water a wet suit is essential, preferably with neoprene a quarter-inch or more in thickness. For casual use, an off-the-rack wet suit works fine but if you spend a lot of time in a wet suit, you will appreciate the comfort of a custom-made one. If you are crawling about in shallow water, knee pads (such as those used by volleyball players) will increase the life of the suit.

If you want to dive using scuba, keep in mind that it is extremely hazardous; small mistakes can result in major injuries or death. Before diving, you should take lessons to become certified; this is required in most states, anyway. You should also be in good health and physical condition. Never dive alone.

PUBLIC AQUARIA

For seeing live fish, a good second choice to diving is a visit to a major public aquarium. Increasingly, these aquaria are setting up tanks that simulate as much as possible natural conditions, down to the beer bottles littering the bottom of exhibits of harbor fishes. Some of the biggest and best public aquaria are:

· Steinhart Aquarium, Golden Gate Park, San Francisco
· Monterey Bay Aquarium, Monterey
· John G. Shedd Aquarium, 1200 S. Lake Shore Drive, Chicago
· The New York Aquarium, Boardwalk and W. 8th Street, Brooklyn
· The New England Aquarium, Central Wharf, Boston
· The National Aquarium, Pier 3, 501 E Pratt Street, Baltimore
· Aquarium of the Americas, French Quarter, New Orleans
· Oregon Coast Aquarium, Newport, Oregon
· Stephen Birch Aquarium Museum, La Jolla, California
· Texas State Aquarium, Corpus Christi

FISH PHOTOGRAPHY

Fish are not easy to photograph. In the water they prefer dimly lit areas and are hard to approach. Out of the water, they flop around, glare in the sun, fold in their fins, and fade rapidly. Still, the astonishing quality of many photographs of fish published in recent years indicates that it is possible to get good pictures. Because I am not a particularly good photographer, I will offer only a few words of advice. For underwater photography, there are a number of specialized cameras available, some of them quite inexpensive (although the quality of photographs may reflect this as well). If you obtain an underwater camera, you should also have a light meter and flash attachment, and be prepared to waste a great deal of film. Out of the natural environment, reasonably good pictures of small fishes can be obtained using an aquarium small enough to be carried out into the field. Have a piece of window glass cut to fit inside the aquarium, so the movements of the subject fish will be restricted to the area in front of the camera. Set the aquarium up so that there is a nondistracting background. If you need to take a quick picture of a fish out of the water, place it on a uniformly colored background, preferably white, and out of the glare of the sun. Most temperate fish can be immobilized temporarily by immersing them in ice water for a few minutes. If the fish to be photographed is killed, the photographs should be taken immediately, before the color changes. On dead fish, the fins can be spread with pins and propped up with modeling clay.

Books on underwater photography can be found in most stores that sell underwater cameras. For obtaining professional-looking photographs of fresh fish, D. D. Flescher provides an excellent guide ("Fish Photography," *Fisheries* 8(4):2–6, 1983). The American Fisheries Society maintains a library of slides of fish that are available for loan and/or duplication. Contact: AFS Photo Library, National Fisheries Center, Route 3, Box 41–Lee Town, Kearneysville, W. Va. 25430 (send an 8½ × 11 envelope with 98¢ postage for a copy of the catalog).

FISH PRINTING

Fish printing, or gyotaku, is an art form developed by the Japanese initially as a means of recording large or unusual fish caught, before the days of refrigeration and taxidermy. It has achieved a small following in recent years in this country, mainly among fish biologists, naturalists, and anglers who appreciate the subject matter. However, it is also being recognized as a legitimate art form and displays by major fish printers in art galleries are not uncommon. Fish printing is appealing to the amateur because reasonably good prints can be produced with minimal skills and materials. Basically, to produce a fish print, a fresh fish is cleaned off and mounted so that the fins are erect. A thin coating of ink is then brushed on with a stiff brush. Newsprint or rice paper is then carefully pressed on the fish to make the print. This is obviously a great oversimplification of the process.

More detailed instructions by one of the nation's leading fish printers, Christopher M. Dewees (one of whose prints appears at the beginning of this chapter), can be found in the pamphlet "Gyotaku: Japanese Fish Printing" (available for $1.25 from Division of Agricultural Sciences, University of California, 1422 S. 10th St., Richmond, Calif. 94804, leaflet 2548). Instructions are also available in Dewees's book *The Printer's Catch: An Artist's Guide to Pacific Coast Edible Marine Animals* (1984, Sea Challengers, 4 Sommerset Rise, Monterey, Calif. 93940). This book contains color plates of fish prints and an informative text on the fishes. The classic book on fish printing is *Gyotaku: The Art and Technique of the Japanese Fish Print* by Yoshio Hiyama (1964, University of Washington Press, Seattle).

FISH ART

Fish photography and fish printing are among the most accessible ways to record the beauty of fish, but sketches made in the field can be delightful and informative records of fishes caught and habitats observed. A fine introduction to sketching is *The Art of Field Sketching* by C. W. Leslie (1984, Prentice-Hall, Englewood Cliffs, N.J.). For more precise drawings, the techniques of scientific illustrators are needed, for which a good introduction is *Scientific Illustration: A Guide for the Beginning Artist* by Z. T. Jastrzebski (1985, Prentice-Hall), who has drawn many fish himself. Paintings of fish by fine artists are part of the growing field of wildlife art and are sometimes featured in the glossy magazine *Wildlife Art News* (P.O. Box 16246, Minneapolis, Minn. 55416-0246). For a history of fish in art see "An Introduction to Fish Imagery in Art" by P. B. Moyle and M. A. Moyle (1991, *Environmental Biology of Fishes* 31:5–23). Subsequent issues of the journal feature articles on individual works by great artists. The first civilization to use fish in art on a regular basis was that of ancient Egypt, featured in *Fish and Fishing in Ancient Egypt* by D. J. Brewer and R. F. Friedman (1989, Aris and Phillips, Warminster, England). The difficulty of separating illustration from art is well shown by the lovely plates in *Classic Natural History Prints: Fish* by S. P. Dance and G. N. Swinney (1990, Arch Cape Press, New York).

Illustration Credits and Acknowledgments

Many thanks to the good judgment and prompt, accurate service of typesetters Pamela Raeke and Trudy Phillips with South/West Printing, Bryan, Texas, who set most of the labels for the black and white illustrations. Thanks also to Rosario Carrizales with Type Direction, McAllen, Texas, who set labels for corrections, additions, and color figures and to Irene Navarro with TypeWorks Typesetting, McAllen, Texas, who set final corrections and changes.

Grateful acknowledgment is given to the following for providing and/or granting permission to use or reprint material. Note: all copyright notices and figure references are written exactly as they appeared in the original sources, except the plate numbers from Jordan publications, which were converted from Roman numeral to Arabic. Full contributor names are provided wherever possible unless the source dictated otherwise. "AND" separates information from separate sources within one figure credit. "Used with permission" statements are included because the permission giver often differed from the original publisher and/or the contributor requested the given format. A bibliography follows the credits and acknowledgments specifics. Length of text and total figure numbers have been omitted from the bibliography because the credit and acknowledgment section refers to specific material used.

Figure 1-1: After photo by E. S. Hobson, plate 24a, photo by Alex Kerstich, plate 24b, between pp. 206 and 207; and drawing by Tor Hansen, figure 80, p. 164 in Thomson, Findley, and Kerstitch 1987. The University of Arizona Press, © 1987. Used with permission of publisher, The University of Arizona Press. AND after photo by Edmund Hobson, Plate 44 in Hobson and Chave 1972. Copyright © 1972 by The University Press of Hawaii. Used with permission of publisher, University of Hawaii Press.

Figure 1-2: [Data] based on pp. 12–15 in Ruttner 1968. Copyright University of Toronto Press, 1963. Used with permission of publisher. AND [graph] based on p. 46 in De Carli 1978. Copyright © 1978 by Arnoldo Mondadori Editore S.p.A., Milan. Used with permission of Arnoldo Mondadori Editore S.p.A., Milan, Italy.

Figure 1-3: [Geologic time scale] after figure 2-5, p. 35 in Hildebrand 1988. Copyright © 1974, 1982, 1988 by John Wiley & Sons, Inc. Used with permission of publisher and author. AND [evolution data] modified from figure 1-1, p. 4 in Moyle and Cech 1982. © 1982 by Prentice-Hall, Inc., Englewood Cliffs, N.J. 07632. AND [extinct fishes] after figure 2-11, p. 23; figure 4-15, p. 74; figure 5-3, p. 87; figure 5-19, p. 100; and figure 8-22, p. 191 in Moy-Thomas and Miles 1971. © 1971 J. Moy-Thomas and R. S. Miles. Adapted with permission of Dr. R. S. Miles.

Figure 1-4: After photo by Gene Wolfsheimer p. 2663 in Burton and Burton, 1970, vol. 20. © 1970 B.P.C. Publishing Limited. Used with permission of Macdonald and Company (Publishers) Ltd., London, England. AND after photo by Dr. Stanislav Frank, figure 119, p. 81 in Frank 1971. © Copyright 1969 Artia, Prague. Used with permission of Dr. Stanislav Frank and Dilia Theatrical and Literary Agency, Prague, Czechoslovakia.

Figure 1-5: After photo by John Bigelow Taylor, plate 13 in Moulard 1984. Photograph Copyright 1981 by John Bigelow Taylor. Plate 13, Copyright 1984 by the Twelvetrees Press. Used with permission of photographer and publisher.

Figure 2-1: [Haida dogfish design ca. 1900, Queen Charlotte Islands] redrawn after pencil sketch by C. F. Newcombe, courtesy Royal British Columbia Museum, Victoria, British Columbia.

Figure 2-2: After specimen courtesy of the Wildlife and Fisheries Museum, University of California, Davis.

Figure 2-3: Modified after drawing by Chris van Dyck, figure 39, p. 127 in Moyle 1976a. Copyright © 1976, by The Regents of the University of California. Modified with permission of publisher, University of California Press, Berkeley.

Figure 2-4: After specimen from public market, Tuxtla Gutierrez, Chiapas, Mexico.

Figure 2-5: After photos by Dr. Stanislav Frank: [mosquito fish] figure 465, p. 304, [sculpin] figure 780, p. 508, [flounder] figure 788, p. 512 in Frank 1971. © Copyright 1969 Artia, Prague. Used with permission of Dr. Stanislav Frank and Dilia Theatrical and Literary Agency, Prague, Czechoslovakia. AND [flounder face] after photo by Jane Burton p. 786, vol. 6; Rat-tailed [rattail] after photo by W. M. Stephens p. 1911, vol. 14; and [swordfish] after drawing by Malcolm McGregor p. 2341, vol. 17 in Burton and Burton 1970, vols. 6, 14, and 17. © 1970 B.P.C. Publishing Limited. Used with permission of Macdonald and Company (Publishers) Ltd., London, England. AND [moray eel] after drawing by Chris van Dyck, p. 43 and [blenny] after photo by Alex Kerstitch, p. 181 in Thomson, Findley, and Kerstitch 1987. The University of Arizona Press, © 1987. Used with permission of publisher, The University of Arizona Press. AND after photos by G. Mazza: [trout] pp. 6–7 and [John Dory] p. 64; and [pike] after photo by Munich Zoo Aquarium, p. 45 in Dareff 1971, vol. 11. © 1970,

1971 Fratelli Fabbri Editori, Milan. Used with permission of Gruppo Editoriale Fabbri S.P.A., Milan, Italy.

Figure 2-6: [Gar-the fish] after photo by Laird Parker, p. 32 in Dareff 1971, vol. 11. © 1970, 1971 Fratelli Fabbri Editori, Milan. Used with permission of Gruppo Editoriale Fabbri S.P.A., Milan, Italy. AND [cross-sections] modified after figure 6-4, p. 99 in Hildebrand 1988. Copyright © 1974, 1982, 1988 by John Wiley & Sons, Inc. Used with permission of publisher and author.

Figure 2-7: After photo by D. A. Thomson, [scorpionfish] figure 37, p. 65; and after photos by Alex Kerstitch: [Pacific spadefish] figure 62, p. 124 and [butterfly fish] figure 67, p. 134 in Thomson, Findley, and Kerstitch 1987. The University of Arizona Press, © 1987. Used with permission of publisher, The University of Arizona Press.

Figure 2-8: [Sole] after photo, figure 7e, p. 806 in Mearns and Sherwood 1974. Used with permission of publisher, Transactions of The American Fisheries Society. AND [black crappie] after photo by Peter B. Moyle.

Figure 2-9: Modification of a photo by Samuel Woo, Illustration Services, University of California, Davis, figure 8.6, p. 105 in Moyle and Cech 1988. © 1988, 1982 by Prentice-Hall, Inc. A Division of Simon & Schuster, Englewood Cliffs, New Jersey 07832.

Figure 3-1: After electron micrograph by M. J. Massingill and S. J. Mitchill. Used courtesy of Joseph J. Cech, Jr., University of California, Davis.

Figure 3-2: After specimen courtesy of the Wildlife and Fisheries Museum, University of California, Davis.

Figure 3-3: After specimen from public market, Tuxtla Gutierrez, Chiapas, Mexico.

Figure 3-4: [Blood cells] after photo, figure 5.3, p. 54 in Moyle and Cech 1988. © 1988, 1982 by Prentice-Hall, Inc. A Division of Simon & Schuster. AND [heart cross-sections] after figure 1, p. 180, in Randall 1968. Copyright © 1968 by the American Zoologist. Used and modified with permission of the American Zoologist and author, D. J. Randall.

Figure 3-5: [Gill filaments on a gill arch] after figure 1, p. 345 in Hughes and Grimstone 1965. Used with permission of Professor G. M. Hughes and the Company of Biologists Limited, Cambridge, England. AND [secondary lamella] in Munshi and Singh 1968. Used with permission B. N. Singh.

Figure 3-6: After specimen courtesy of the Wildlife and Fisheries Museum, University of California, Davis.

Figure 4-1: After photo plate 27, p. 203 in Kroeber and Barrett 1960. Copyright 1960 University of California Press. Used with permission of University of California Press, Berkeley.

Figure 4-2: [Spawning ground data] from figure 39, p. 131 in Harden Jones 1968. Used with permission of Dr. F. R. Harden Jones. AND [spawning adult, newly hatched larvae, and food items] after figures 27a, 27b, and 27c, p. 65 in Muus and Dahlstrom 1971. © EDICIONES OMEGA, S.A.–Barcelona, 1971. Used with permission of G·E·C GADS FORLAG, Copenhagen, Denmark.

Figure 4-3: [Enlarged fish] after photo by D. P. Wilson, p. 1374 in Burton and Burton 1970, vol. 10. © 1970 B.P.C. Publishing Limited. Used with permission of Macdonald and Company (Publishers) Ltd., London, England.

Figure 4-4: [Broodhider] after two photos by K. Paysan, p. 40, and figure 3a, p. 41 in *Aquarium Digest International*, issue no. 22 (English Translation of TI 40). Used with permission of publisher, TetraWerke. AND [scatterer] after photo by William Pflieger, p. 1 and p. 3 in Pflieger and Belusz 1982. 1982 © Missouri Conservation Commission. Used with permission of the Missouri Conservation Commission. AND [external bearers] after figure 46, p. 93 in Muus and Dahlstrom 1971. © EDICIONES OMEGA, S.A. – Barcelona, 1971. Used with permission of G·E·C GADS FORLAG, Copenhagen, Denmark. AND [internal bearers and guarder] after photos by Peter B. Moyle.

Figure 4-5: After observations of live specimens in Russian River, California.

Figure 5-1: Adapted from illustration by Richard Ellis, plate 5, pp. 56–57 in Ellis 1983. © 1983 by Richard Ellis. Reprinted by permission of the artist. AND [pygmy goby] after specimen courtesy of the Department of Ichthyology, California Academy of Sciences, San Francisco, California.

Figure 5-2: [Adult, sucking disk] after figures 1a and 1b, p. 33 in Muus and Dahlstrom 1971. © EDICIONES OMEGA, S.A. – Barcelona, 1971. Used with permission of G·E·C GADS FORLAG, Copenhagen, Denmark. AND [adult attached to host fish] after photo, figure 7, p. 148 in Hardisty and Potter 1971. Copyright © 1971 by Academic Press Inc. Ltd. Used with permission of publisher, Academic Press Inc., Orlando, Florida. AND [nest] redrawn after figure 3, p. 292 in Lohnisky 1966. Used with permission of publisher, The Journal of Czechoslovak Zoological Society, Prague, Czechoslovakia. AND [mating pair] redrawn after figure 50, p. 324 in Sterba 1962. Used with permission of Guenther Sterba. AND [hagfish mouth detail] after photo p. 86 in Jensen 1966. Copyright © 1966 by Scientific American, Inc. All rights reserved. Used with permission of Scientific American, Inc. and David Jensen. AND [hagfish] after specimen courtesy Wildlife and Fisheries Museum, University of California, Davis.

Figure 5-3: [Smooth dogfish, cow shark, and spiny dogfish] and [stingray] after photos by Chris Mari van Dyck, courtesy Steinhart and Monterey Bay Aquariums. AND [sixgill shark] and [guitarfish] after photos by Peter B. Moyle. AND [angle shark] after photo by Napoli Aquarium, p. 123 (bottom) and [manta ray] after photo by Milan Museum of Natural History, p. 141 in Dareff 1971, vol. 11. © 1970, 1971 Fratelli Fabbri editori, Milan. Used with permission of Gruppo Editoriale Fabbri S.P.A., Milan, Italy. AND [sawfish] after photo by Marineland of Florida. Used by permission of Marineland of Florida. AND [ratfish] after photo © 1972 Western Marine Laboratory, Santa Barbara, California. Used by permission William C. Jorgensen.

Figure 5-4: [Whole specimen] redrawn after drawing by D. Bryan Stone III, figure, p. 68 in Castro 1983. Copyright © 1983 José I. Castro. AND [mouth detail] after photo by L. J. V. Compagno, figure 15.3, p. 210 in Moyle and Cech 1988. © 1988, 1982 by Prentice-Hall, Inc. A Division of Simon and Schuster. AND [damage detail] after photo by D. Castleberry, courtesy of Peter B. Moyle.

Figure 5-5: [Lungfish] after live specimen courtesy Steinhart Aquarium. AND [coelacanth] after drawing, p. 220 in McClane 1974. Copyright © 1965 Holt, Rinehart & Winston, Inc. AND Copyright © 1974 by A. J. McClane.

Figure 5-6: [Beluga] after photo by Dr. Stanislav Frank, figure 55, p. 37 in

Frank 1971. © Copyright 1969 Artia, Prague. Used with permission of Dr. Stanislav Frank and Dilia Theatrical and Literary Agency.

Figure 5-7: [Moray eel] after photos by Chris Mari van Dyck, courtesy of Steinhart Aquarium. AND [snipe eel] redrawn after figure 156, plate 54, and [gulper eel] redrawn after figure 175, plate 56 in Jordan and Evermann 1900. AND [garden eel] after photo by Alex Kerstitch, plate 2d, between p. 78 and p. 79 in Thomson, Findley, and Kerstitch 1987. The University of Arizona Press, © 1987. Used with permission of publisher, The University of Arizona Press. AND [European eel, adult] redrawn after figure, p. 139 in De Carli 1978. Copyright © 1975 by Arnoldo Mondadori Editore S.p.A., Milan. English Translation Copyright © 1978 by Arnoldo Mondadori Editore S.p.A., Milan. Used with permission of Arnoldo Mondadori Editore S.p.A., Milan, Italy. AND [European eel, development] redrawn from figure 42e, p. 83 in Muus and Dahlstrom 1971. © EDICIONES OMEGA, S.A.–Barcelona, 1971. Used with permission of G·E·C GADS FORLAG, Copenhagen, Denmark.

Figure 5-8: After live specimen, courtesy of Steinhart Aquarium.

Figure 5-9: [Sonar ship] redrawn from figure 15.2, p. 348 in Longwell, Flint, and Sanders 1969. Copyright © 1969 by John Wiley & Sons, Inc. Reprinted by permission of John Wiley & Sons, Inc. AND [bristlemouths] redrawn from drawings by Malcolm McGregor, p. 283 in Burton and Burton 1970, vol. 10. © 1969 B.P.C. Publishing Limited. Used with permission of Macdonald and Company (Publishers) Ltd., London, England.

Figure 5-10: [Mexican tetra] after figure 179, p. 147 in Sterba 1963. © Vista Books, Longacre Press Ltd., London, 1962. Used with permission of Guenther Sterba. AND [minnow, carp, and sucker] after specimens courtesy Wildlife and Fisheries Museum, University of California, Davis.

Figure 5-11: [Bristlemouth catfish] after photo by Dr. W. Foersch, p. 18 in Andrews n.d.*b*. Used with permission of publisher, TetraWerke. AND [armored catfish] after photos by K. Paysan, cover, p. 6 and p. 7 in Rakowicz et al. n.d. Used with permission of publisher, TetraWerke. AND [bony-plated catfish] after live specimen courtesy Steinhart Aquarium. AND [brown bullhead] after specimen courtesy of Wildlife and Fisheries Museum, University of California, Davis. AND [white, blind, cave-inhabiting catfish] redrawn from drawings by Carol Fletcher Daniels, figure 1, p. 3 and figure 2 (bottom), p. 4 in Lundberg 1982. Used with permission of John G. Lundberg and publisher, University of Michigan.

Figure 5-12: [Lanternfish] after figure 247, plate 92; AND [hake] after figure 884, plate 359 and [grenadier] after figure 914, plate 369 in Jordan and Evermann 1900. AND [anglerfish] modified from figure 3, plate 10 in Regan and Trewavas 1932. AND [frogfish] redrawn from drawing by Tor Hansen, p. 53 in Thomson, Findley, and Kerstitch 1987. The University of Arizona Press, © 1987. Used with permission of publisher, The University of Arizona Press.

Figure 5-13: [Flying fish] after figure 318, plate 118 and [hatchetfish] after figure 261, plate 97 in Jordan and Evermann 1900.

Figure 5-14: After specimen courtesy Wildlife and Fisheries Museum, University of California, Davis.

Figure 5-15: [Sea horse] after live specimen courtesy Steinhart Aquarium.

AND [pipefish and tubesnout] after specimens courtesy Wildlife and Fisheries Museum, University of California, Davis. AND [stickleback] after photo by Peter B. Moyle.

Figure 5-16: [Black and yellow rockfish, yellowtail rockfish, and blue rockfish] after photo by Peter B. Moyle. AND [copper rockfish and kelp rockfish] after photos by Chris Mari van Dyck, courtesy Monterey Bay Aquarium.

Figure 5-17: [Yellowfin goby] after drawing by Alan Marciochi, figure 39, p. 348 in Moyle 1976a. Copyright © 1976, by The Regents of the University of California. Used with permission of publisher, University of California Press, Berkeley. AND [mudskipper] after photos by Jane Burton: Photo Res p. 1526 and p. 1527 in Burton and Burton 1970, vol. 11. © 1970 B.P.C. Publishing Limited. Used with permission of Macdonald and Company (Publishers Ltd., London, England.

Figure 5-18: After specimens from public market, Tuxtla Gutierrez, Chiapas, Mexico.

Figure 5-19: Modified after drawings by Betsy B. Washington, figure 2, p. 406; figure 3, p. 407; figure 4, p. 408; and figure 5, p. 409 in Richardson, Dunn, and Naplin 1980. Used with permission of authors, Jean R. Dunn and Nancy Anne Naplin.

Figure 5-20: After photos by Marineland of Florida. Used with permission of Marineland of Florida.

Figure 6-1: After film by Ron and Rose Eastman. Eastman and Eastman 1969. Used with permission of Rose Eastman.

Figure 6-2: [Temperature requirement data] from figure 5, p. 25 in Mihursky and Kennedy 1967. Used with permission of publisher, Transactions of the American Fisheries Society. AND [egg, alevin, and fry] after photos in Haig-Brown 1967. © Roger Duhamel, F.R.S.C., Queen's Printer and Controller of Stationery, Ottawa, 1967. Used with permission of publisher, Information and Publications Branch, Department of Fisheries and Oceans, Ottawa. AND [parr and adult] after drawing by Chris van Dyck, figure 35, p. 117 in Moyle 1976a. Copyright © 1976, by The Regents of the University of California. Used with permission of publisher, University of California Press, Berkeley.

Figure 6-3: [Graph] after figure 2, p. 86 in Elston and Bachen 1976. Used with permission of publisher, Transactions of the American Fisheries Society. AND [fish] redrawn from drawing by Alan Marciochi, figure 99, p. 272 in Moyle 1976a. Copyright © 1976, by The Regents of the University of California. Used with permission of publisher, University of California Press, Berkeley.

Figure 6-4: [Pearlfish] after photo by William M. Stephens, p. 1718 in Burton and Burton 1970, vol. 13. © 1970 B.P.C. Publishing Limited. Used with permission of Macdonald and Company (Publishers) Ltd., London, England. AND [sea cucumber] adapted from figure 19-41, p. 773 in Barnes 1974. © 1974 by W. B. Saunders Company. Copyright 1963 and 1968 by W. B. Saunders Company.

Figure 7-1: [Lure] after photo p. 435 in McClane 1974. Copyright © 1965 Holt, Rinehart & Winston, Inc. and Copyright © 1974 by A. J. McClane. AND [brook trout] after drawing by Chris van Dyck, figure 46, p. 148; and [sucker and sculpin] after drawings by Alan Marciochi: figure 17(b), p. 79, figure 78,

p. 225, and figure 127, p. 354 in Moyle 1976*a*. Copyright © 1976, by The Regents of the University of California. Used with permission of publisher, University of California Press, Berkeley. AND [dace] after specimen courtesy Wildlife and Fisheries Museum, University of California, Davis.

Figure 7-2: [Blackfly larva] modified after figure 14.19a, p. 402; and [nonbiting midge larva] after figure 14.22i, p. 413 in Usinger et al. 1963. Copyright 1956 by The Regents of the University of California. Used with permission of publisher, University of California Press, Berkeley. AND [stonefly larva, dragonfly larva, bottom of submerged rock, and caddisfly larva] after specimens collected and photos by Chris Mari van Dyck. AND [caddisfly larva in net] modified after figure 65, p. 236 in Ruttner 1968. English translation of Third Edition Copyright University of Toronto Press, 1963. Used with permission of publisher, University of Toronto Press, Buffalo, New York.

Figures 7-3 and 7-4: Photos by Peter B. Moyle.

Figure 7-5: [Top] data from table 1, p. 182; [bottom] data from table 4, p. 185 in Moyle 1976*b*. Used with permission of publisher, California Fish and Game.

Figure 7-6: [Water snake] copied from negative #334945 of the Photographic Collection of the American Museum of Natural History. Used courtesy Department of Library Services, American Museum of Natural History. AND [kingfisher] after photo by Ian Tait, plate 433 in Bull and Farrand 1977. All rights reserved. Copyright 1977 under the International Union for the protection of literary and artistic works (Berne). Used with permission of photographer, Ian C. Tait.

Figure 7-7: After photos by Charles R. Demas, figure 2, p. 6, and figure 3, p. 7 in Demas 1973.

Figure 8-1: After photo by John Lazenby, p. 42 in Palmer 1980. Copyright © 1980 The Pennsylvania State University.

Figure 8-2: [Map] modified after figure 1, p. 3; [data for chart] after table 1, pp. 16 and 17, and table 2, p. 18 in Smith and Powell 1971. Used with permission of The American Museum of Natural History.

Figure 8-3: [Stoneroller] after specimen courtesy R. A. Daniels and The New York State Museum, Albany, New York. AND [dace, bluegill, and smallmouth bass] after specimens courtesy of Wildlife and Fisheries Museum, University of California, Davis.

Figure 9-1: [Mountain whitefish] drawing by Chris van Dyck, figure 32, p. 109 in Moyle 1976*a*. Copyright © 1976, by The Regents of the University of California. Used with permission of publisher, University of California Press, Berkeley. AND after paintings by Duane Raver, [walleye] no. 26 and [yellow perch] no. 25 from series of 30 illustrations published by the U.S. Government, Department of Interior, U.S. Fish and Wildlife Service. AND after photos by Douglas R. Stamm, [northern pike] p. 17 and [muskellunge] p. 15 in Stamm 1977. Copyright © 1977 The Regents of the University of Wisconsin System. Used with permission of publisher, The University of Wisconsin Press. AND [northern pike] after photo by Munich Zoo Aquarium, p. 45 (top) in Dareff 1971, vol. 11. © 1970, 1971 Fratelli Fabbri Editori, Milan. Used with permission of Gruppo Editoriale Fabbri S.P.A., Milan, Italy.

Figure 9-2: [Cool-water lake and riverine lake] photos by Evelyn W. Moyle. Used with permission of photographer. AND [cold-water alpine lake and California reservoir] photos by Peter B. Moyle.

Figure 9-3: Drawings by Chris van Dyck, figure 42 (top), p. 137, and figure 47, p. 152 in Moyle 1976a. Copyright © 1976, by The Regents of the University of California. Used with permission of publisher, University of California Press, Berkeley.

Figure 9-4: Drawing by Chris van Dyck, figure 5, p. 137 in Moyle 1976a; modified after Miller 1951; and using Kokanee data from Cordone et al. 1971. Copyright © 1976, by The Regents of the University of California. Used with permission of publisher, University of California Press, Berkeley. AND figure 28.4, p. 389 in Moyle and Cech 1988. © 1988, 1982 by Prentice-Hall, Inc. A Division of Simon & Schuster.

Figure 9-5: Figure 28.3, p. 387 in Moyle and Cech 1988. © 1988, 1982 by Prentice-Hall, Inc. A Division of Simon & Schuster.

Figure 9-6: After drawings by Alan Marciochi, [suckers] figure 72, p. 213; [catfish] figure 86, p. 242; and [threadfin shad] figure 30, p. 101 in Moyle 1976a. Copyright © 1976, by The Regents of the University of California. Used with permission of publisher, University of California Press, Berkeley. AND [white bass] after painting by Duane Raver, no. 13 from series of 30 illustrations published by the U.S. Government, Department of Interior, U.S. Fish and Wildlife Service. AND [carp] after photo, p. 51 (bottom) in Dareff 1971, vol. 11. © 1970, 1971 Fratelli Fabbri Editori, Milan. Used with permission of Gruppo Editoriale Fabbri S.P.A., Milan, Italy.

Figure 10-1: [Damselfly] after photo by Chris Mari van Dyck. AND after paintings by Duane Raver: [largemouth bass] no. 22, [redear sunfish] no. 20, and [channel catfish] no. 10 from a series, illustrations published by the U.S. Government, Department of Interior, U.S. Fish and Wildlife Service. AND [bluegill] after specimen courtesy of Wildlife and Fisheries Museum, University of California, Davis. AND [fathead minnow] after drawing by Alan Marciochi, figure 67, p. 200 in Moyle 1976a. Copyright © 1976, by The Regents of the University of California. Used with permission of publisher, University of California Press, Berkeley.

Figure 10-2: Photo by Chris Mari van Dyck.

Figure 11-1: [Estuary] after photo by R. G. Schmidt, BSFW. AND [delta smelt] after drawing by Alan Marciochi, figure 48, p. 156 in Moyle 1976a. Copyright © 1976, by The Regents of the University of California. Used with permission of publisher, University of California Press, Berkeley. AND [white perch] redrawn after figure 479, plate 181 in Jordan and Evermann 1900. AND [spotted seatrout] after painting by Duane Raver, no. 29 from a series of 30 illustrations published by the U.S. Government, Department of Interior, U.S. Fish and Wildlife Service.

Figure 11-2: Data after table 31-1, p. 417 in Moyle and Cech 1988. © 1988, 1982 by Prentice-Hall, Inc. A Division of Simon & Schuster.

Figure 11-3: [Striped bass] after painting by Duane Raver, no. 14 from a series of 30 illustrations published by the U.S. Government, Department of Interior, U.S. Fish and Wildlife Service. AND [opossum shrimp] after photo by

Jane Burton: Photo Res, p. 1619 in Burton and Burton 1970, vol. 12. © 1970 B.P.C. Publishing Limited. Used with permission of Macdonald and Company (Publishers) Ltd., London, England.

Figure 12-1: After photo by L. T. Findley, figure 5, p. 8 in Thomson, Findley, and Kerstitch 1987. The University of Arizona Press, © 1987. Used with permission of publisher, The University of Arizona Press.

Figure 12-2: [Small fish] after drawing by Tor Hansen, p. 226 in Thomson, Findley, and Kerstitch 1987. The University of Arizona, © 1987. Used with permission of publisher, The University of Arizona Press.

Figure 12-3: [Prickleback] modified from plate 341 in Jordan and Evermann 1900. AND [blenny] modified after photo by P. Scoones, plate 96, and [gunnel] modified after photo by C. C. Hemmings, plate 120 in Lythgoe and Lythgoe 1975. Copyright © 1971 by Blandford Press Ltd. Used with permission of publisher, Doubleday a DIVISION of Bantum, Doubleday, Dell Publishing Group, Inc.

Figure 12-4: After photos by Stan Hess and B. A. Woodling, p. 973 in Burton and Burton 1970, vol. 7. © 1970 B.P.C. Publishing Limited. Used with permission of Macdonald and Company (Publishers) Ltd., London, England.

Figure 12-5: After drawing by Alan Marciochi, figure 124, p. 346 in Moyle 1976a. Copyright © 1976, by The Regents of the University of California. Used with permission of publisher, University of California Press, Berkeley.

Figure 12-6: [Saltmarsh] after photo at South Padre Island, Texas, by Chris Mari van Dyck. AND [mummichog] redrawn after figure 273, plate 102 in Jordan and Evermann 1900.

Figure 12-7: After photo by R. G. Schmidt, BSFW. AND [snook] after figure 476, plate 179; [tarpon] after figure 177, plate 67 in Jordan and Evermann 1900.

Figure 13-1: [Kelp, otter, senorita, opaleye, halfmoon, surfperch, and bass] after photos by Chris Mari van Dyck, courtesy Monterey Bay Aquarium. AND after photos by Alex Kerstitch: [bass] figure 46, p. 84 and [California sheephead] figure 76, p. 150 in Thomson, Findley, and Kerstitch 1987. The University of Arizona Press, © 1987. Used with permission of publisher, The University of Arizona Press.

Figure 13-2: [Sucking disk detail] after photo by John Tashjian, Shedd Aquarium, p. 1942; and [shark with attached remora] after photo by Photo Library Inc., p. 1943 in Burton and Burton 1970, vol. 14. © 1970 B.P.C. Publishing Limited. Used with permission of Macdonald and Company (Publishers) Ltd., London, England.

Figure 14-1: After photos by Alex Kerstitch: [grouper] figure 46, p. 84; and [damselfish] plate 17e, between p. 206 and p. 207; and after drawing by Alex Kerstitch, [damselfish] figure p. 143 in Thomson, Findley, and Kerstitch 1987. The University of Arizona Press, © 1987. Used with permission of publisher, The University of Arizona Press. AND [damselfish] redrawn after figure 594, plate 236 in Jordan and Evermann 1900. AND after photos by Edmund Hobson: [goatfish] plate 1, p. 29, and [trumpetfish] plate 7, p. 18 in Hobson and Chave 1972. Copyright © 1972 by The University Press of Hawaii. Used with permission of University of Hawaii Press.

Figure 14-2: Photo courtesy Steinhart Aquarium, San Francisco, California.
Figure 14-3: After photo p. 1708 in Burton and Burton 1970, vol. 13. ©
1970 B.P.C. Publishing Limited. Used with permission of Macdonald and Company (Publishers) Ltd., London, England.
Figure 15-1: Photo courtesy The Oakland Museum, Kahn Collection.
Figure 15-2: [Nile perch] numerous sources. AND [haplochromine cichlids]
after photos in Ribbink et al. 1983.
Figure 16-1: Fish print by Christopher Dewees.

The plates were done primarily but in most cases not exclusively from
sources listed.
Plate 1: [High Rock Spring and tui chub] after photos by T. Ford, July 1977.
AND [tilapia] after photo by Glenn Black, April 1975, and photo by W. Hauser.
Plate 2: [Grand Canyon] after photo by Chris Mari van Dyck. AND after
photos by J. N. Rinne, [humpback chub] plate 2 (4), p. 14; [bonytail chub] plate
2 (6), p. 14; and [Colorado squawfish] plate 5 (8), p. 17 in Johnson 1987. AND
[razorback sucker] after photo by Nevada Department of Fish and Game.
Plate 3: [Map] after figure 1.(a), p. 3 in Scoppettone et al. 1986. AND [Pyramid Lake, fishway to Marble Bluff Dam, Derby Dam, and cui-ui] after photos
by Peter B. Moyle. AND [fishladder and stripping female cutthroat trout] after
photos by RGB, 816 N. Highland Ave., Hollywood, Ca. 90038. Used courtesy
of Gary Scoppettone.
Plate 4: [Map] after map by University of Wisconsin Sea Grant Institute/
Jana Fothergill, p. 3, and [ciscos with scars] after photo by U.S. Fish and Wildlife Service, p. 4 in University of Wisconsin Sea Grant Institute 1982. Copyright
1982 Board of Regents–University of Wisconsin System–Sea Grant Institute.
AND after drawings by Charles B. Hudson, [blackfin] plate 4 and [bloater] plate
5 in Jordon and Evermann 1911. AND [alewife] after cover photo in Smith
1968. AND [lakefront scene and sailing ship] after photos WHi(X3)12856 and
WHi(X3)30573 from State Historical Society of Wisconsin. Used with permission of Iconographic Collections of the State Historical Society of Wisconsin.
Plate 5: [Owens Native Fish Sanctuary] after photo by E. P. Pister, November
1977. AND [underwater scene and male pupfish] after photo by Thomas L.
Taylor, May 1981. AND [female pupfish] after specimen courtesy Wildlife and
Fisheries Museum, University of California, Davis. AND after drawings by
Carol Mortensen, [map] figure 2, p. 4 and [Owens chub] figure 20, p. 29;
[speckled dace] after photo by Alan Heller, figure 19, p. 28; and [Owens sucker]
after photo by E. P. Pister, figure 22, p. 29 in Soltz and Naiman 1978. ALSO
[Owens chub] after photo by L. Fisk, plate 2, p. 14 in Johnson 1987. AND
[releasing pupfish into sanctuary] after photo by David L. Soltz, August 1970.
Courtesy of Phil Pister.
Plate 6: [Map] after map, [dam] after photo AV3637-1, and [snail darter]
after photos N77.001, N.75-027, N76-019 courtesy Tennessee Valley Authority. ALSO [snail darter] after photo by E. C. Thompson. Copyright March 27,
1977, The Atlanta Journal and Constitution Magazine.
Plate 7: After photos by Peter B. Moyle.

Plate 8: [Virgin River and spinedace] after photos courtesy of Richard A. Heckmann, Brigham Young University. AND [scolex] after photo by Richard A. Heckmann, figure 1, p. 227 in Heckmann, Greger, and Deacon 1987. © American Society of Parasitologists 1987. Used with permission of authors and the American Society of Parasitologists. AND [red shiners] after photos by Ken Bouc, cover and p. 64 in Bouc 1987. Copyright 1987 by the Nebraska Game and Parks Commission all rights reserved. AND [grass carp] after photo by Dan Guravich, p. 49 in Moyle 1984. ALSO [grass carp] after drawing, figure 258, p. 84 in Eddy and Underhill 1978.

Bibliography for Illustrations

Andrews, C., ed. n.d.*a*. *Aquarium Digest International*. Issue no. 22 (English translation by Geoffrey Burwell of TI 40). West Germany: TetraWerke.

Andrews, C., ed. n.d.*b*. *Aquarium Digest International*. Issue no. 27 (English translation by Geoffrey Burwell of TI 49). West Germany: TetraWerke.

Barnes, R. D. 1974. *Invertebrate Zoology*. 3d ed. Philadelphia: W. B. Saunders.

Bouc, K. 1987. *The Fish Book*. Lincoln: Nebraska Game and Parks Commission.

Bull, J., and J. Farrand, Jr. 1977. *The Audubon Society Field Guide to North American Birds: Eastern Region*. New York: Chanticleer Press.

Burton, M., and R. Burton, eds. 1970. *The International Wildlife Encyclopedia*. Vols. 3, 6, 7, 10–15, 17, 19, 20. New York: Marshall Cavendish.

Castro, J. 1983. *Sharks of North American Waters*. College Station, Texas: The Texas A&M Press.

Cordone, A., S. Nicola, P. Baker, and T. Frantz. 1971. "The Kokanee Salmon in Lake Tahoe." *Calif. Fish Game* 57:28–43.

Dareff, H., ed. 1971. *The Illustrated Encyclopedia of the Animal Kingdom*. Vol. 11. New York: Danbury Press.

De Carli, F. 1978. *The World of Fish*. New York: Gallery Books.

Demas, C. R. 1973. "Food Habits of the Rainbow Trout in Relation to the Local Biota of Lower Hat Creek." Unpublished M.S. thesis, California State University, Humboldt.

Eastman, R., and R. Eastman. 1969. *The Private Life of the Kingfisher*. Color, 16 min., sd 16 mm, (BBC;FI); (ISBN 0-699-23861-71).

Eddy, S., and J. C. Underhill. 1978. *How to Know the Freshwater Fishes*. Dubuque, Iowa: Wm. C. Brown Co.

Ellis, R. 1983. *The Book of Sharks*. New York: Harcourt Brace Jovanich.

Elston, R., and B. Bachen. 1976. "Diel Feeding Cycle and Some Effects of Light on Feeding Intensity of the Mississippi Silverside, *Menidia audens*." *Trans. Amer. Fish. Soc.* 105:84–88.

Frank, S. 1971. *The Pictorial Encyclopedia of Fishes.* New York: Hamlyn.

Haig-Brown, R. L. 1967. *Canada's Pacific Salmon.* Ottawa, Canada: Dept. of Fisheries and Oceans.

Harden Jones, F. R. 1968. *Fish Migration.* London: Edward Arnold.

Hardisty, M. W., and T. C. Potter, eds. 1971. *The Biology of Lampreys*, vol. 1. London: Academic Press.

Heckmann, R. A., P. D. Greger, and J. E. Deacon. 1987. "New Host Records for the Asian Fish Tapeworm, *Bothriocephalus acheilognathi*, in Endangered Fish Species from the Virgin River, Utah, Nevada, and Arizona." *J. Parasit.* 73:226–227.

Hildebrand, M. 1988. *Analysis of Vertebrate Structure.* New York: John Wiley.

Hobson, E. S., and E. H. Chave. 1972. *Hawaiian Reef Animals.* Honolulu: University of Hawaii Press.

Hughes, G. M., and A. V. Grimstone. 1965. "The Fine Structure of the Secondary Lamellae of the Gills of *Gadus pollachius*." *Quart. J. Micro. Sci.* 106:343–353.

Jensen, D. 1966. "The Hagfish." *Sci. Amer.* 214:82–90.

Johnson, J. E. 1987. *Protected Fishes of the United States and Canada.* Bethesda, Maryland: American Fisheries Society.

Jordan, D. S., and E. C. Starks. 1895. "The Fishes of Puget Sound." *Proc. Calif. Acad. Sci.* 5:785–855.

Jordan, D. S., and B. W. Evermann. 1900. *The Fishes of North and Middle America.* Washington, D.C. Bull. U.S. Nat. Mus. 47.

Jordan, D. S., and B. W. Evermann. 1911. *Review of the Salmonoid Fishes of the Great Lakes, with Notes on the Whitefishes from Other Regions.* Washington, D.C. Bull. U.S. Bur. Fish. 29.

Kroeber, A. L., and S. A. Barrett. 1960. *Fishing among the Indians of Northwestern California, with Special Data from E. L. Gifford and G. W. Hewes.* Berkeley and Los Angeles: University of California Press.

Lohnisky, K. 1966. "The Spawning Behaviour of the Brook Lamprey, *Lampetra planeri*." *Vestnik Ceskosiovenske Spolecnosti Zoologicke* 30:289–307.

Longwell, C. R., R. F. Flint, and J. E. Sanders. 1969. *Physical Geology.* New York: John Wiley.

Lundberg, J. G. 1982. "The Comparative Anatomy of the Toothless Blindcat, *Trogloglanis pattersoni* Eigenmann, with Phylogenetic Analysis of Ictalurid Catfishes." Univ. Mich. Mus. Zool. Misc. Publ. 163.

Lythgoe, J., and G. Lythgoe. 1975. *Fishes of the Sea.* New York: Doubleday.

McClane, A. J., ed. 1974. *McClane's New Standard Fishing Encyclopedia and International Angling Guide.* New York: Holt, Rinehart, and Winston.

Mearns, A. J., and M. Sherwood. 1974. "Environmental Aspects of Fin Erosion and Tumors in Southern California Dover Sole." *Trans. Amer. Fish. Soc.* 4:799–810.

Mihursky, J. A., and V. S. Kennedy. 1967. "Water Temperature Criteria to Protect Aquatic Life." In *A Symposium on Water Quality Criteria to Protect Aquatic Life*, ed. E. L. Cooper, 20–32. Amer. Fish. Soc. Spec. Publ. 4.

Miller, R. G. 1951. "The Natural History of Tahoe Fishes." Unpublished Ph.D. dissertation, Stanford University.

Mouland, B. L. 1984. *Within the Underworld Sky: Mimbres Ceramic Art in Context*. Pasadena, Calif.: Twelvetrees Press.

Moyle, P. B. 1969. "Ecology of the Fishes of a Minnesota Lake, with Special Reference to the Cyrpinidae." Unpublished Ph.D. dissertation, University of Minnesota.

Moyle, P. B. 1976*a*. *Inland Fishes of California*. Berkeley, Los Angeles, London: University of California Press.

Moyle, P. B. 1976*b*. "Some Effects of Channelization of the Fishes and Invertebrates of Rush Creek. Modoc County, California." *Calif. Fish Game* 62:179–186.

Moyle, P. B. 1984. "America's Carp." *Natural History* 93:42–51.

Moyle, P. B., and J. J. Cech, Jr. 1982. *Fishes: An Introduction to Ichthyology*. 1st ed. Englewood Cliffs, N. J.: Prentice-Hall.

Moyle, P. B., and J. J. Cech, Jr. 1988. *Fishes: An Introduction to Icthyology*. 2d ed. Englewood Cliffs, N.J.: Prentice-Hall.

Moy-Thomas, J. A., and R. S. Miles. 1971. *Paleozoic Fishes*. Philadelphia: W. B. Saunders.

Munshi, J. D., and B. N. Singh. 1968. "A Study of Gill-epithelium of Certain Freshwater Teleostean Fishes with Special Reference to the Air-breathing Fishes." *J. Zool. India* 9:91–107.

Muus, B. J., and P. Dahlstrom. 1971. *Guia de los peces de mar del Atlantico y del Mediterraneo: pesca–biologia, importancia economica*. Barcelona, Spain: Ediciones Omega, S.A.

Palmer, T. 1980. *Rivers of Pennsylvania*. University Park, Pennsylvania: Penn State Press.

Pflieger, W., and L. C. Belusz. 1982. *An Introduction to Missouri Fishes*. Missouri Conservation Commission.

Rakowicz, M., et al., eds. n.d. *Aquarium Digest International*. Issue no. 21 (English translation by Geoffrey Burwell). West Germany: TetraWerke.

Randall, D. J. 1968. "Functional Morphology of the Heart of Fishes." *Amer. Zool.* 8:179–189.

Raver, D. No date. *A Series of 30 Original Paintings*. Washington, D.C.: U.S. Fish and Wildlife Service.

Regan, C. T., and E. Trewavas. 1932. "Deep-sea Angler Fishes (Ceratoidea)." *Dana Report 2.*

Ribbink, A. J., B. A. Marsh, A. C. Ribbink, and B. J. Sharp. 1983. "A Preliminary Survey of the Cichlid Fishes of the Rocky Habitats in Lake Malawi." *S. Afr. J. Zool.* 18:149–310.

Richardson, S. L., J. R. Dunn, and N. A. Naplin. 1980. "Eggs and Larvae of Buttersole, *Isopsetta isolepis* (Pleuronectidae), off Oregon and Washington." *Fishery Bull.* 78:401–418.

Ruttner, F. 1968. *Fundamentals of Limnology*. 3d ed. Canada: University of Toronto Press.

Scoppettone, G. G., M. Coleman, and G. A. Wedemeyer. 1986. *Life History and Status of the Endangered Cui-ui of Pyramid Lake, Nevada.* Washington, D.C.: U.S. Fish and Wildlife Service.

Smith, C. L., and C. R. Powell. 1971. "The Summer Fish Communities of Brier Creek, Marshall County, Oklahoma." *American Museum Novitiates* 2458:1–30.

Smith, S. H. 1968. "The Alewife." *Limnos* 1:8–12.

Soltz, D. L. and R. J. Naiman. 1978. *The Natural History of Native Fishes of the Death Valley System.* Los Angeles: Natural History Museum, Los Angeles County.

Stamm, D. R. 1977. *Underwater: The Northern Lakes.* Madison: University of Wisconsin Press.

Sterba, G. 1962. *Vie Neunaugen (Petromyzonidae) Handbuch der Binnenfischerei Mitteleuropas.* Vol. 3. Stuttgart: E. Schweizerbart.

Sterba, G. 1963. *Freshwater Fishes of the World.* London: Vista Books.

Thomson, D. A., L. T. Findley, and A. N. Kerstitch. 1987. *Reef Fishes of the Sea of Cortez: The Rocky-shore Fishes of the Gulf of California.* Tucson: The University of Arizona Press.

University of Wisconsin Sea Grant Institute. 1982. "The Sea Lamprey: Invader of the Great Lakes." *The Great Lakes Alien Series No. 1.* Madison: University of Wisconsin Sea Grant Institute.

Usinger, R. L., et al., eds. 1963. *Aquatic Insects of California with Keys to North American Genera and California Species.* Berkeley, Los Angeles: University of California Press.

Williams, J. D., and D. K. Finnley. 1977. "Our Vanishing Fishes: Can They Be Saved?" *Frontiers* 1977:35–41.

Index

Designer: U.C. Press Staff
Compositor: Wilsted & Taylor
Text: 10/13 Sabon
Display: Sabon
Printer: Thomson-Shore, Inc.
Binder: Thomson-Shore, Inc.